TRANSGENIC FISH

TRANSGENIC FISH

edited by

Choy L. Hew
Research Institute, Hospital for Sick Children and
Departments of Clinical Biochemistry and Biochemistry
University of Toronto
Toronto, Ontario,
Canada

Garth L. Fletcher
Ocean Science Centre and
Department of Biology
Memorial University of Newfoundland
St. John's, Newfoundland,
Canada

World Scientific
Singapore • New Jersey • London • Hong Kong

Published by

World Scientific Publishing Co. Pte. Ltd.
5 Toh Tuck Link, Singapore 596224
USA office: 27 Warren Street, Suite 401-402, Hackensack, NJ 07601
UK office: 57 Shelton Street, Covent Garden, London WC2H 9HE

Library of Congress Cataloging-in-Publication Data
Transgenic fish / edited by Choy L. Hew, Garth L. Fletcher.
p. cm.
Includes bibliographical references and index.
ISBN-13 978-981-02-0997-1 -- ISBN-10 981-02-0997-5
1. Transgenic fish. I. Hew, Choy L. (Choy Leong), 1942–
II. Fletcher, G. L.
QL638.99.T73 1992
597'.015--dc20 92-1256
CIP

British Library Cataloguing-in-Publication Data
A catalogue record for this book is available from the British Library.

Cover photograph: Shows the dramatic size difference between a transgenic Atlantic salmon and some negative controls. The transgenic salmon was produced using an antifreeze protein promoter linked to a salmon growth hormone gene. The controls were also injected, but not transgenic (see Du et al., this monograph).

The photograph was taken by John Bourne of Memorial University of Newfoundland.

PREFACE

The production of transgenic animals is now established as a powerful technique that can be utilized for basic as well as applied research. Transgenic mice, for example, are making significant contributions to our understanding of gene expression, the regulation of cellular function and embryonic development. In addition, they also serve as essential models for studying immunology, cancer and other genetic disorders in humans. Transgenic technology is also being used to improve the economics of animal production for food by improving for example, growth efficiencies, growth rates and disease resistance.

In contrast to mammals, studies on transgenic fish are still very much in their infancy. Despite this, it is already evident that such fish will be of considerable economic benefit to aquaculture. Furthermore, it is apparent that transgenic fish also enable us to make quantum leaps in our understanding of physiological and biochemical mechanisms unique to fish, and of the developmental biology of vertebrates in general.

Since the field of transgenic fish is relatively new its potential for research and industrial development is only just beginning to be widely recognized. Therefore it is likely that many scientists are contemplating exploring the merits of this technology for the first time. With this in mind we decided that it would be valuable to put together a monograph that encompassed the full spectrum of current research on transgenic fish.

The monograph is divided into three sessions. The first session on oogenesis, models, gene transfer and detection techniques opens with a brief description of the embryology of a model teleost, the desert pupfish. This is followed by two chapters (chapters 2 and 3) on the various approaches in gene transfer which includes direct injection, injection via the micropyle and electroporation. Chapter 4 and to a lesser extent, chapter 10 includes details of polymerase chain reaction strategies used for the detection of transgenesis, a technique which has greatly enhanced the sensitivity and ease of detection. Chapter 5 illustrates the attractive features of the Japanese medaka as a model fish for transgenic studies. Although not included in this monograph, the zebrafish shares many common features with the medaka. The second session focuses on transgene inheritance and expression. Chapters 6 to 10 summarize recent research results from a wide variety of fish species with major emphasis on growth hormone gene transfer. The growth rate of many of these transgenic fish are dramatic, ranging from a modest 20-40% up to several fold increase in size. Chapter 11 is a summary for antifreeze protein gene transfer. The antifreeze protein and growth hormone are both important polypeptides with well defined biological functions that are directly relevant to aquaculture. It is therefore not surprising that this is reflected in many contributions in this monograph.

The farming of transgenic fish has raised many questions concerning their ecological impact due to accidental release. The final two chapters discuss these issues in terms of safety precautions, guidelines and methods of containment. Several approaches to sterilize transgenic fish are suggested.

As indicated above, research on transgenic fish has only just begun in earnest, thus it is clear that the experiments we see being performed today can only be regarded as a "first generation" approach. There is little doubt that future studies will increase in their sophistication using more efficient gene transfer methods and better gene constructs that are highly specific for the desired trait capable of exogenous regulation and the precise integration site by gene targeting. This will require considerable research into the molecular biology of fish in order to identify appropriate promoters, enhancers, silencers and tissue specific elements and to determine the full effects of the gene product on the organism. Exogenous control over a transgene in free-living fish may be difficult, but it could be accomplished by manipulating diet, photoperiod, temperature, salinity or utilizing responses elicited by the fishes' chemosensory system.

ACKNOWLEDGMENT

We thank Linda Gardiner for typing and editing the monograph. We also thank the Medical Research Council, the Natural Sciences and Engineering Research Council, the Canadian National Centres of Excellence Program (Ocean Production Enhancement Network) and the Department of Fisheries and Oceans, Government of Canada for their generous funding support of our research on transgenic fish.

In memory of

KEVIN SCOTT GUISE

February 7, 1950 - February 12, 1991

This monograph on Transgenic Fish is dedicated to the memory of Professor Kevin Scott Guise, an early pioneer in research on transgenic fish. Kevin was a professor at the University of Minnesota in the Department of Animal Science when he succumbed to illnesses associated with his hemophilia. In his tragically short career Kevin had established an international reputation in the application of molecular genetics to analyzing and modifying the genomes of fish and poultry.

Kevin was raised in Kettering, Ohio, received his B.S. degree, Summa cum Laude, in Botany from the University of Rhode Island in 1972, and took his M. Phil. (1978) and Ph.D. (1981) in Biology/Somatic Cell Genetics at Yale University where he worked with Professor Frank Ruddle. Dr.Guise continued his training as a molecular biologist as a postdoctoral fellow in Prof. R. A. Gravel's laboratory in the Genetics Department, Hospital for Sick Children, Toronto, Canada from 1981 to 1984. He joined the Department of Animal Sciences at the University of Minnesota in 1984 as an Assistant Professor to work on the transfer of growth-promoting genes into chickens.

In 1985 Kevin and three of his colleagues at the University of Minnesota, Tony Faras, Perry Hackett, and Anne Kapuscinski, formed a collaboration to investigate the possibilities of using transgenic technology in fish. The three molecular biologists (Kevin, Tony and Perry) had a lot of ideas about the possibilities of altering fish phenotypes, but no real idea of how to keep fish alive and what fish were likely candidates for genetic alteration; Anne Kapuscinski provided the critical addition to the team that came to be known (first in grant application) as the Minnesota Transgenic Fish Group (MTFG). Owing to his enthusiasm for communicating with other molecular biologists and researchers in fisheries, Kevin was selected as the leader of the MTFG. Between 1985 and 1990 Kevin participated in several national and international meetings where transgenic fish research got its beginnings and published more than 20 papers on transgenic fish and its implications.

Professor Guise's research in both poultry and fish was oriented towards elucidating the basic molecular genetics of growth promoting genes in fish and poultry, developing methods of gene transfer into recipient animal eggs, and mapping genes to specific chromosomes. His lab used molecular techniques to identify and characterize fish stocks (and populations) for both wildlife management and commercial applications. Within the MTFG, Dr. Guise was responsible for leading the group in the isolation and characterization of growth hormone genes (primarily with Drs. John Schneider and Tony Faras), development of piscine expression vectors (with Drs. Zhanjiang Liu and Perry Hackett), gene transfer (with postdoctoral fellows Eric Hallerman and John Schneider) and maintenance of transgenic fish under extraordinary conditions (primarily with Mark Gross and Dr. Anne Kapuscinski). In addition, Dr. Guise helped to bring several distinguished foreign visitors to spend sabbaticals with the MTFG, including Professors Zuoyan Zhu (then at the Institute of Hydrobiology in Wuhan, China, presently at the Dept. of Zoology, University of Aberdeen, Scotland), Boaz and Neomi Moav (Dept. of Zoology, Tel Aviv University, Israel), and M. Carmen Alvarez (Dept. of Cell Biology and Genetics, University of Malaga, Spain). Kevin also set up a critical collaboration with Professor Choy Hew (Hospital for Sick Children, Toronto, Canada) which allowed the MTFG to transfer "all-fish" recombinant DNA's, containing Hew's chinook salmon growth hormone sequence, to fertilized fish eggs. During the period of 1986-1988 approximately 200,000 eggs from walleye, northern pike, Atlantic salmon, and rainbow trout were microinjected with a variety of recombinant DNAs harboring either growth hormone gene sequences or reporter genes such as the bacterial chloramphenicol acetyl transferase. Additionally, an enormous number of goldfish and zebrafish were microinjected during preliminary experiments with various recombinant DNAs. Dr. Guise's lab concentrated primarily on developing methods of gene transfer to fish. His first Ph.D. student in Genetics, Dr. Sung-Joo Yoon, in 1986 examined DNA-binding to sperm as one possible method of mass delivery before she went on to investigate the possibilities of using co-transfer of the neomycin phosphotransferase (neo) gene as a selectable marker in transgenic fish. Two contributions in this monograph review a portion of his work.

Kevin was afflicted with hemophilia. As a child he was confined to a wheelchair. His mother would wheel him through the parks during the warm months and cultivate his interests indoors with nature and biology books. Later Kevin was able to walk with the help of a cane. Consequently, one of the biggest thrills of his life was to be featured in Sports Illustrated in a story covering genetic manipulation of fish to create superfish for sportsfishing. The problems associated with the hemophilia occasionally slowed Kevin's movements, but never his eagerness and enthusiasm for his research activities in both fish and poultry. Just as much of his experimental work in transgenic fish was reaching fruition, infections took his life. His legacy is a richer environment for all who come into the field of transgenic fish in particular and transgenic animals in general. We will miss his contributions at so many levels.

CONTRIBUTORS

Dr. Tom Chen
Institute of Marine Biotechnology
University of Maryland
Baltimore, Maryland, USA

Dr. Peter L. Davies
Biochemistry Department
Queen's University
Kingston, Ontario, Canada

Robert H. Devlin
West Vancouver Laboratory
Fisheries and Oceans Canada
West Vancouver, British Columbia, Canada

Dr. Edward Donaldson
West Vancouver Laboratory
Fisheries and Oceans Canada
West Vancouver, British Columbia, Canada

Shao Jun Du
Department of Biochemistry
University of Toronto
Toronto, Ontario, Canada

Rex A. Dunham
Department of Fisheries and Allied Aquaculture
Auburn University
Auburn, Alabama, USA

Anthony J. Faras
Department of Microbiology and Institute of Human Genetics
University of Minnesota
St. Paul; Minnesota, USA

Dr. Garth Fletcher
Marine Sciences Research Laboratory
Ocean Sciences Centre
Memorial University of NFLD
St. John's, Newfoundland, Canada

Sally V. Goddard
Marine Sciences Research Laboratory
Ocean Sciences Centre
Memorial University of NFLD
St. John's, Newfoundland, Canada

Dr. Zhiyuan Gong
Departments of Biochemistry and Clinical Biochemistry,
University of Toronto
Toronto, Ontario, Canada

Mark L. Gross
Department of Fisheries and Wildlife
University of Minnesota
St. Paul, Minnesota, USA

Dr. Kevin Guise
Department of Animal Science and Institute of Human Genetics
University of Minnesota
St. Paul, Minnesota, USA

Perry Hackett
Genetics and Cell Biology and Institute of Human Genetics
University of Minnesota
St. Paul, Minnesota, USA

Eric M. Hallerman
Department of Fisheries and Wildlife Sciences
Virginia Polytechnic Institute and State University
Blacksburg, Virginia, USA

Dr. Choy L. Hew
Departments of Biochemistry and Clinical Biochemistry,
University of Toronto
Toronto, Ontario, Canada

Koji Inoue
Central Research Laboratory
Nippon Suisan Kaisha
Hachioji, Japan

Dr. Anne R. Kapuscinski
Department of Fisheries and Wildlife
University of Minnesota
St. Paul, Minnesota, USA

Madonna J. King
Ocean Sciences Centre
Memorial University of NFLD
St. John's, Newfoundland, Canada

Chun-Mean Lin
Centre of Marine Biotechnology
University of Maryland and
Department of Biological Sciences
University of Maryland Baltimore
County
Baltimore, Maryland, USA

Zhanjiang Liu
Department of Genetics and Cell
Biology
University of Minnesota
St. Paul, Minnesota, USA

Dr. Boaz Moav
George S. Wise Faculty of Life
Sciences
Department of Zoology
Tel Aviv University
Ramat Aviv
Tel Aviv, Israel

Neomi L. Moav
Department of Animal Science
University of Minnesota
St. Paul, Minnesota, USA

Dr. K. Ozato
Department of Biology
Faculty of Liberal Arts and Sciences
Kyoto University
Kyoto, Japan

Dennis A. Powers
Hopkins Marine Station
Stanford University
Pacific Grove, California, USA

Margaret A. Shears
Marine Sciences Research Laboratory
Ocean Sciences Centre
Memorial University of NFLD
St. John's, Newfoundland, Canada

Dr. Jurgen R. Vielkind
Departments of Endocrinology and
Pathology, B.C. Cancer Research
Centre and University of BC
Vancouver, British Columbia,
Canada

Yuko Wakamatsu
Department of Biology,
Faculty of Liberal Arts and Sciences
Kyoto University
Kyoto, Japan

Dr. Mel Weisbart
Department of Biology
University of Regina
Regina, Saskatchewan, Canada

Dr. Zuoyan Zhu
Institute of Marine Biotechnology
University of Maryland
Baltimore, Maryland, USA

Institute of Hydrobiology
Chinese Academy of Sciences
Wuhan, Hubei, P.R. China

Department of Zoology
University of Aberdeen
Aberdeen, Scotland

CONTENTS

CHAPTER 1

Egg Development: From Gametes to Embryonic Development and Hatching

Melvin Weisbart

ABSTRACT

The formation of gametes is described with emphasis on oogenesis including vitellogenesis and the reinitiation of meiosis and the maturation of the oocyte. The development of the fertilized fish egg is described in 40 stages for the desert pupfish, *Cyprinodon macularius*.

FORMATION OF GAMETES

Oogenesis

The formation of mature oocytes from oogonia has been described for the rainbow trout, *Oncorhynchus mykiss* (Yamamoto et al., 1965) and for the Japanese medaka, *Oryzias latipes* (Yamamoto, 1963, 1964). These stages include the mitotic division of oogonia into primary oocytes which undergo growth and development during which a chorion forms between the plasma membrane of the oocyte and the follicle cells which surround the oocyte. The primary oocyte is arrested in prophase I of meiosis. During this period of arrest vitellogenin is synthesized in the liver under estrogen stimulation, secreted into the plasma and probably transported into rainbow trout oocytes as vitellogenin by endocytosis (Tyler et al., 1988). The vitellogenin is then broken down to lipovitellin and phosvitin which remain in the yolk. The completion of this vitellogenic growth of the oocyte is followed by the release of the maturation-inducing steroid (MIS) which has been shown to be 17α,20ß-dihydroxy-4-pregnen-3-one (DHP) in teleosts (Nagahama, 1987; Scott and Canario, 1987) and 17α,20ß,21-trihydroxy-4-pregnen-3-one in the Atlantic croaker, *Micropogonias undulatus* (Trant et al., 1986; Trant and Thomas, 1989). The synthesis of DHP requires the presence of both the thecal and granulosa layers of the oocyte follicle (Young et al., 1986). The DHP acts at the plasma membrane of the oocyte by binding to a membrane bound receptor (Maneckjee et al. 1989, 1991). The interaction of DHP with its oocyte membrane receptor stimulates the formation of maturation-promotion factor (MPF) which leads to the reinitiation of meiosis to complete germinal vesicle breakdown (GVBD) and the formation and extrusion of the first polar body (Nagahama, 1987) and formation of the secondary oocyte (haploid). The completion of meiosis requires the penetration of a spermatozoon into the secondary oocyte which results in the formation and extrusion of the second polar body and ultimate fusion of the pronuclei.

Spermatogenesis

The formation of mature spermatozoa is a complicated process beginning in the mitosis of spermatogonia (diploid) and ending with the formation of mature spermatozoa (haploid) which are capable of fertilizing the mature ovum. The process of mitosis and meiosis involved in forming mature spermatozoa has been reviewed by Nagahama (1983) and includes presumably important residence in the vas deferens just as spermatozoa in mammals must mature in the epididymis after leaving the contacts with the Sertoli cell.

EMBRYONIC DEVELOPMENT

The importance of describing the stages of embryonic development of fish has been well established in the literature. The normal stages of the killifish, *Fundulus heteroclitus* (Oppenheimer, 1937) have been used by many investigators undertaking experimental studies in embryonic development. For example, Solberg (1938) studied the effect of x-radiation on different stages of killifish embryos. Extensive research in Japan and elsewhere has been performed on embryonic development of the Japanese medaka, *Oryzias latipes* (Ando, 1962) using the stages described for the medaka (Matsui, 1949).

Having established some of the later stages of development for lake trout embryos, Garside (1959) investigated the time required to reach certain of these stages under conditions of low oxygen levels. In addition, normal stages of salmon development (Pelluet, 1944) have been used by Hayes and Pelluet (1945) and Hayes et al. (1953) as a basis for studying environmental effects of temperature changes on embryos; similar studies also have been carried out for the Ludoga whitefish, *Coregonus lavaretus ludoga* (Vernidub, 1956). Other teleosts in which staging of the developing embryo have been described include the platyfish, *Platypoecilus maculatus* (Tavolga, 1947), the chum salmon, *Oncorhynchus keta* (Mahon and Hoar, 1956), the zebrafish, *Brachydanio rerio* (Hisaoka and Battle, 1958), the blue gourami, *Trichogaster trichopterus* (Hisaoka and Firlit, 1962). Recently, the zebrafish has been heralded as an excellent model for studying development (Barinaga, 1990, Ho and Kane, 1990).

The desert pupfish, *Cyprinodon macularius*, a cyprinodont, follows the general developmental pattern of other cyprinodonts such as the killifish and the Japanese medaka. Nevertheless, the stages of development described in what follows gives more stages of development that are observable in vivo without killing the embryo.

The pupfish belongs to the order Cyprinodontiformes, suborder Cyprinodontoidei, superfamily Cyprinodontoidae (Nikol'skii, 1961). Members of the superfamily Cyprinodontoidae generally inhabit brackish water. In 1853, Baird and Girard (1853) were the first investigators to describe and name the species *Cyprinodon macularius*. Later, various investigators described what appeared to be other related species, namely *C. californiensis*, *C. browni*, and *C. nevadensis*. More recently, all of these species were considered synonymous with *C. macularius*. In 1943, Miller (1943) showed that *C. macularius* and *C. nevadensis* were in fact separate species.

Desert pupfish are restricted to the basin of the lower Colorado River and the Sonoyta River of northern Sonora, Mexico (Miller, 1948). They are found in shore pools, 5 to 15 metres wide and from 20 to 200 metres in length, along the shores of the Salton Sea in California (Barlow, 1958). This habitat is of particular interest because these pools are very shallow, being about 10 cm or less in depth; diurnal temperature fluctuations may be as great as 20^{0}C, and salinities may go as high as $80\ {}^{0}/_{00}$ to $90\ {}^{0}/_{00}$. Nevertheless, these fish have the capacity to survive these conditions (Barlow, 1958a), and to spawn in salinities as high as $70\ {}^{0}/_{00}$ with successful development (Kinne and Kinne, 1962).

The teleost egg is surrounded by an enveloping membrane often referred to as the chorion which is elevated at fertilization resulting in the formation of a perivitelline space. "A true chorion ... is formed as a gelatinous or filamentous layer produced external to the radiate membrane by the follicle cells" (Nelson, 1953). Since the enveloping membrane of killifish consists of two parts, a "chorion internum" which is the transformed vitelline membrane or zona radiata of the oocyte plus a villous layer which is the "chorion externum" and is synthesized by the follicle cells (Kemp, personal communication to Kaighn (Nelson, 1953)), the killifish membrane is not a true chorion, but it has been referred as such. Although no information is available concerning the origin of the pupfish membrane, it is probable the term chorion is also incorrectly applied to the pupfish membrane since it is very similar to that of the killifish.

Stages of normal embryological development of the desert pupfish

The description of the embryonic development of the desert pupfish will be largely based on morphological and physiological features that can readily be observed under the phase microscope in the living egg. The time (in hours:minutes) after fertilization (A.F.) to reach a certain stage is given for eggs developing in dechlorinated fresh water (Toronto tap water) under 100% air saturation of the water at 28 ± 0.3°C with pH between 8.2 and 8.5 and no photoperiod.

Stage 1: unfertilized ovum (Fig.1) The yolk of the unfertilized egg of the pupfish is surrounded by a thin surface of protoplasm which is directly beneath the non-elevated chorion. The presumptive animal pole is indicated by the micropyle of the chorion. Presumably, the egg pronucleus lies in the superficial protoplasmic layer directly beneath the micropyle as it is in the medaka (Sakai, 1961). Small highly refractive oil globules are dispersed over the yolk. Between the two poles there is a large central oil globule. Occasionally, however, there may be two or three central oil globules close together. The micropyle persists up to five hours if the ovum has not been fertilized, but whether the egg remains competent has not been investigated. Generally most unfertilized eggs become activated either due to handling or when in contact with water (Zotin, 1958). These eggs invariably can be handled without crushing, whereas the unfertilized ova readily collapse. The term activation is used here to signify the breakdown of cortical alveoli, subsequent formation of a perivitelline space and closure of the micropyle. Activation may or may not be followed by blastodisc formation.

The chorion of the pupfish consists of at least three layers: an outer villous layer by which the demersal egg is anchored to a substrate, a middle layer and an inner fibrous layer which appears to consist of several laminae. The middle layer of the killifish egg is not digested by the action of chorionase (Kaighn, 1964).

The egg is bright yellow and slightly opaque due to the presence of cortical alveoli in the thin peripheral layer of the ovum. The unfertilized ovum including the chorion measures 1.35 mm in diameter (average of ten eggs) at this time.

Stage 2: one-celled egg (Fig. 2) Upon fertilization, the cortical alveoli disappear with a subsequent increased clarity of the yolk and formation of a perivitelline space due to the separation of plasma membrane and chorion. A lenticular blastodisc of finely granular cytoplasm, manifests itself by 0:15 A.F.; this is found under the remnants of the micropyle. By 0:25 A.F. a thin cap of cortical cytoplasm has been raised up from the yolk and at 0:45 A.F. one cell can be seen. The diameter of the yolk is 1.33 mm.

Stage 3: two-celled egg at 0:55 A.F. (Fig. 3) Cleavages of the blastodisc are typically meroblastic. The first cleavage is meridional and divides the blastodisc into two cells of about equal size.

Stage 4: four-celled egg at 1:25 A.F. (Fig. 4) The second cleavage planes are not meridional but longitudinal and occur at right angles to the first division resulting in four cells of unequal size; two large cells and two small cells. The large cells are found diagonally across from one another as are the small ones.

Stage 5: eight-celled egg at 1:50 A.F. (Fig. 5) The third cleavage planes are parallel to the first and produce eight cells of unequal size at the surface. The four corner cells are larger than the remaining four cells.

Stage 6: sixteen-celled egg at 2:10 A.F. (Fig. 6) The fourth cleavage furrows are longitudinal and parallel to the second cleavage. They occur synchronously thereby producing 16 cells.

Stage 7: thirty-two-celled egg at 2:30 A.F. (Fig. 7) The fifth cleavage divides the blastoderm into two layers of cells. The cleavage can best be seen under the microscope by observing the formation of two layers of cells among the central cells. Beyond this stage it is difficult to follow subsequent cleavages without resorting to histologically prepared serial sections.

Stage 8: early high blastula at 3:30 A.F. (Fig. 8) Continuation of mitosis after the thirty-two-celled stage results in a blastoderm which is raised high off the yolk; it consists of small cells approximately 0.08 mm in diameter. The small oil globules, which were scattered over the yolk at fertilization, are now in a dense clump at the vegetal pole. The blastoderm measures 0.66 mm in diameter and 0.30 mm in height from the yolk. No periblast nuclei are visible.

Stage 9: late high blastula at 4:00 A.F. (Fig. 9) Although the blastoderm remains raised high from the yolk, the cells are smaller in size; measuring approximately 0.06 mm in diameter. The blastoderm measures 0.68 mm in diameter and 0.32 mm in height from the yolk. Nuclei from the peripheral cells have migrated into the periblast, establishing at least one row of periblast nuclei. The highly refractive oil globules originally dispersed over the yolk are now aggregated into a dense ball at the vegetal pole.

Stage 10: intermediate blastula at 5:00 A.F. (Fig. 10) As the blastoderm begins to spread over the yolk, the diameter increases to 0.71 mm and the height from the yolk decreases greatly. The cells continue to divide and are about 0.03 mm in diameter. There are approximately 4 rows of periblast nuclei.

Stage 11: flat blastula at 6:00 A.F. (Fig. 11) The blastoderm sinks further into the yolk until the plasma membrane of the yolk and the enveloping layer of blastoderm cells form a continuously smooth circle in lateral view. The blastoderm is no longer raised from the yolk. The diameter of the blastoderm is 0.71 mm. There are about 6 rows of periblast nuclei present, but the number is not constant around the blastoderm.

Stage 12: early gastrula at 9:00 A.F. (Fig. 12) The blastoderm continues to spread over the yolk and its diameter is 0.91 mm. There are approximately 9 rows of periblast nuclei presumably as a result of mitosis of the peripheral cells of the blastoderm (Wilson, 1889). The central portion of the blastoderm is partly thinned out while the periphery is thickened to form the germ ring. There is a thickening at one point of the germ ring called the embryonic shield. The future embryo will form within this shield. The embryonic shield extends apically from the edge of the blastoderm 0.30 mm.

Stage 13: epiboly covering 1/3 yolk at 11:00 A.F. (Fig. 13) The blastoderm has covered one third the diameter of the yolk. The central portion of the blastoderm is very thin while the germ ring and embryonic shield are considerably thickened. There are only about 2 rows of periblast nuclei observable.

Stage 14: middle gastrula at 12:00 A.F. (Fig. 14) The blastoderm covers half the yolk. A dorsal view of the embryonic shield does not reveal any condensation of cells to form the neural keel, but a lateral view of the shield indicates a darkened strip of cells. These cells make up the presumptive central nerve (usually nervous but it only becomes "nervous" with inputs) system. No periblast nuclei are evident.

Stage 15: epiboly covering 2/3 yolk at 14:00 A.F. (Fig. 15) When the blastoderm covers two thirds of the yolk, the embryonic shield clearly manifests two parallel lines which diverge slightly at the animal pole and represent the anlage of the central nerve system. The embryonic axis is thereby established in an animal-vegetal pole direction.

Stage 16: late gastrula at and formation of optic anlagen at 15:00 AF. (Fig. 16) The blastoderm continues to move towards the vegetal pole of the egg by epiboly. The optic anlagen form at the neural keel and close to the animal pole of the egg thereby establishing the anterior-posterior directions of the embryo. The outpocketing of the optic primordium is a solid mass of cells rather than vesicular.

Stage 17: formation of 2 pairs of somites at 18:00 A.F. (Fig. 17) In the posterior region of the lateral mesoderm, which can be clearly seen on either side of the central nerve system, 2 pairs of somites are formed. The germ ring which is continuous with the mesoderm has almost reached the vegetal pole; a small blastopore therefore exists. The presumptive heart tissue is clearly visible but does not manifest any differentiation. The optic chiasma are well defined. Although no pigmentation has occurred, chromatoblasts are observable over the yolk surface. The embryo's length is more than one third of the yolk circumference. A small refractile vesicle is apparent just under the tail. It is probably not Kupffer's vesicle but rather the yolk vesicle described for the stone cat, *Notorus* (Sumner, 1900) and for the flying fish, *Hirundichthys affinis* (Evans, 1961). The presumptive brain tissue is not differentiated.

Stage 18: formation of auditory placodes and closure of the blastopore at 19:00 A.F. (Fig. 18) The yolk is covered completely with cells from the blastoderm. The number of somite pairs has increased to four. The auditory placodes appear on either side of the central nerve system; the brain has not yet differentiated into the three primary segments.

Stage 19: formation of optic vesicles at 20:00 A.F. (Fig. 19) The presumptive eye tissue, which was represented by a solid evagination of cells from the neural keel at 15:00 A.F. is now vesicular. A narrow slit forming the optocoele is clearly evident. Although the heart is not readily observable in all embryos, some show definite morphological indications that it is forming. The large central oil globule very often interferes with and prevents accurate observations of heart formation. There are at this time 6 pairs of somites. The mesencephalon and rhombencephalon are beginning to differentiate.

Stage 20: early pigmentation at 21:00 A.F. (Fig. 20) Pale brown pigmentation can be observed on the embryo, the yolk sac epithelium and the heart. The number of somites has increased to 7 pairs.

Stage 21: formation of optic cups at 22:00 A.F. (Fig. 21) Tubulation of the presumptive forebrain, midbrain and hindbrain to form a neurocoele is in very early stage, for only a very thin slit is observable in the nerve tissue. The optic vesicles have begun to invaginate away from the lateral ectoderm to form the optic cups. Presumably the lens placodes, which are now formed on the overlying ectoderm, are the result of induction by the optic vesicles.

Body chromatophores and heart pigmentation have increased in intensity to a mid-brown colour. Ectoderm overlying the 8 or 9 pairs of somites is pigmented as well. This type of chromatophore has been designated as the "guanophore" (Hishida et al., 1961). It has also been referred by Goodrich (1927) as the brown chromatophore.

Stage 22: formation of auditory vesicles at 24:00 A.F. (Fig. 22) The auditory vesicles are formed but no otoliths are visible. The mesocoele is established in the mesencephalon. The optic lobes, which are formed, span 0.13 mm. There are 11 to 12 pairs of somites organized. Pigmentation has intensified to a dark brown, both on the embryo and on the yolk.

Stage 23: expansion of optic lobes 26:00 A.F. (Fig. 23) The mesocoele in the mesencephalon continues to increase in size resulting in a span of 0.23 mm of the optic lobes. The metencephalon is beginning to differentiate. The embryo, which surrounds more than one half the yolk circumference, has 15 pairs of somites.

Stage 24: initiation of heart beat at 28:00 A.F. (Fig. 24) Concomitant with the expansion of the optic lobes to a diameter of 0.29 mm is the differentiation of the metencephalon. Blood vessels can be seen forming in the yolk sac. The heart is beginning to beat but no blood is circulating. The olfactory placodes have formed and there are 17 pairs of somites.

<u>Stage 25: increase in heart rate at 30:00 A.F. (Fig. 25)</u> The number of somites has increased to 20. The optic lobes and the eyes span 0.30 mm and 0.37 mm respectively. The heart contracts at the rate of about 45 beats per minute but no blood cells can be seen circulating.

<u>Stage 26: oscillation of blood cells in heart at 32:00 A.F. (Fig.26)</u> The heart rate has increased to approximately 60 beats per minute and blood cells can be seen to oscillate within the heart tube. The optic lobes have expanded to 0.33 mm while the span of the eyes has increased to 0.38 mm. The yolk vesicle is reduced greatly in size. There are about 21 pairs of somites.

<u>Stage 27: initiation of blood circulation at 34:00 A.F. (Fig. 27)</u> The optic lobes and eyes have increased their span to 0.36 and 0.42 mm respectively, while the heart rate has increased to approximately 85 beats per minute. Blood cells can be observed flowing sequentially through the differentiating regions of the heart and ventral aorta and on through the embryo by way of the aortic arches (at least two arches are present). The aortic arches have fused to form the dorsal aorta at about the level of the presumptive pectoral fins. The blood continues to flow through the dorsal aorta and into the caudal artery in the tail region of the embryo. The tail, which has lifted off the yolk, is short. Blood in the tail is collected into the ventral caudal vein. It passes anteriorly to the trunk region of the embryo where it flows into the posterior vitelline vein which bifurcates in the yolk sac, sending blood to the left and to the right of the embryo. The third aortic arches can also be seen coming from the ventral aorta and going into the head. The left and right anterior cardinal veins drain blood from the head passing it posteriorly to the left and right anterior vitelline veins (ducts of Cuvier). Occasionally the left or right anterior vitelline vein has been observed not to pass blood freely; instead the vein is filled with oscillating blood cells. If this situation persists, the embryo generally fails to successfully develop to hatching.

<u>Stage 28: pectoral fin bud formation at 38:00 A.F. (Fig. 28)</u> The presumptive pectoral fin tissue observable at 34:00 A.F. has now formed a distinct mass. The distance between the tips of the two masses, that is, the span, is 0.40 mm. The yolk mass has decreased to 1.3 mm in diameter when measured at right angles to the embryo. The caudal artery does not extend to the very tip of the tail but reverses its direction a short distance from the tail tip giving rise to the ventral caudal vein described above.

However, for a short distance from the end of the caudal artery there is a short posterior extension filled with oscillating blood cells. The heart is no longer a straight tube due to the anterior movement of the ventricle; the atrium and ventricle form an "L"-shaped tube.

Stage 29: aggregation of otoliths at 41:00 A.F. (Fig. 29) The scattered otoliths in the auditory vesicles have undergone concretion into larger grains which aggregate into two clumps of otoliths. One otolith is small and located ventro-anteriorly, while the other is small and located dorso-posteriorly in the auditory vesicle. Blood cells can be seen to oscillate within the blood vessels found in the intersegmental tissue, that is, the tissue between two somites.

Stage 30: blood flow in the dorsal intersegmental arterioles at 43:00 A.F. (Fig. 30) Blood flow can be observed in the dorsal intersegmental arterioles associated with the anterior body somites; there is oscillation of cells in the middle vessels and there are no moving blood cells associated with the posterior somites. The pectoral fins, which form just posterior to the ducts of Cuvier, span 0.42 mm. Just posterior to the left posterior vitelline vein, a small blood vessel has formed with few, if any, blood cells flowing through it.

Stage 31: early retinal pigmentation at 47:00 A.F. (Fig. 31) Pigmentation of the retina is beginning. The retinal cells are outlined by thin rims of pigment. A few small black pigment cells are present at the junction of the telencephalon and diencephalon. Similar cells are also visible laterally near the rhombencephalon. These chromatophores are black when viewed with either transmitted or reflected light. Embryos fixed with G-Bouins solution retain the black pigment. When fully developed these cells are the largest and most conspicuous type of chromatophore. Melanophores of the Japanese medaka have been described which are large, most conspicuous and insoluble in alcohol (Goodrich, 1927). This information suggests that the chromatophores described above for the desert pupfish are melanophores.

Stage 32: intermediate retinal pigmentation at 50:00 A.F. (Fig. 32) The retinal cells are surrounded by thin rims of pigment which appear black under both transmitted and reflected light. Melanophores on the head and trunk have expanded in size. The yolk diameter at right angles to the body axis is 1.22 mm and the span of the pectoral fins is 0.47 mm.

Stage 33: late retinal pigmentation at 54:00 A.F. (Fig. 33) Individual retinal cells are surrounded by thick rims of black pigment, and the retinal pigmentation pattern is beginning to appear. Melanophores extend about half way down the dorsal surface of the tail and they have increased in number and size. The spans of the yolk and pectoral fins are 1.19 and 0.50 mm respectively.

Stage 34: establishment of retinal pigment patterns at 63:00 A.F. (Fig. 34) A distinct pattern has formed on the retina; this is visible only by reflected light. A few short lines of black pigment surround the lens. Lines of black pigment radiate from this incomplete circle of pigment to the periphery of the retina. Guanophores at the base of the caudal fin have begun to migrate to the periphery of the fin.

Stage 35: blood circulation in pectoral fins at 68:00 A.F. (Fig. 35) The pectoral fins, which have been seen to move sporadically and contain oscillating blood cells since 66:00 A.F., now have blood circulating through the peripheral blood vessel. The pectoral fins span 0.70 mm; the yolk is 1.08 mm in diameter. The guanophores in the caudal fin have migrated further from the base.

Stage 36: development of dorsal fin rays in the caudal fin at 72:00 A.F. (Fig. 36) Dorsal fin rays can be seen forming in the caudal fin. The caudal artery ramifies in the caudal fin and sends arterioles along one side of the developing rays and on into the periphery of the fin. The blood from the fin flows back into the caudal vein by way of venules which travel along the opposite side of the fin rays. There are approximately 4 to 5 pairs of these fin ray blood vessels.

Stage 37: mouth open at 78:00 A.F. (Fig. 37) The mouth is formed and open as well as capable of occasional movement. The yolk diameter and pectoral fin span are 1.02 and 0.75 mm respectively. The hatching glands can be seen forming in the lining of the oral cavity and the inner lining of the opercula.

Stage 38: formation of central caudal vein at 92:00 A.F. (Fig. 38) The central caudal vein extends from near the base of the caudal fin to the yolk; here it runs diagonally away from the dorsal aorta into the yolk circulation. The ventral caudal vein continues to function. The caudal artery and central caudal vein are enclosed within the hemal arches of the tail vertebrae. Most of the chromatophores on the yolk sac have fused and the vitelline blood can be seen flowing through channels formed beneath these pigment bridges. There are 8 caudal fin rays present. Melanophores on the head are large; individual cells are up to 0.06 mm in diameter. The hatching glands appear fully formed and measure approximately 0.02 mm in diameter; they are distributed as described above.

<u>Stage 39: hatching at 108:00 A.F. (Fig. 39)</u> Just prior to hatching the mouth, opercula and pectoral fins can be seen moving very rapidly. The perivitelline fluid becomes somewhat milky, indicating the release of the hatching enzyme from the hatching glands. Approximately 7 minutes later the embryo, by means of a series of somersaults which result in the tail striking the chorion, breaks the chorion. The tail is the first to penetrate the chorion; then with violent body movements the embryo frees itself completely from the chorion (outer and middle layers of which remains undigested). The yolk sac measures 0.80 mm in diameter. The total length of the prolarva measured from the anterior tip of the premaxillaries to the end of the caudal fin, is 4.67 mm. The caudal fin has 8 dermal fin rays, two of which are in early development. There are 17 post-anal vertebrae formed, and one additional vertebra is forming. The ventral caudal vein has begun to atrophy and the blood flow in the central caudal vein has increased. The time of hatching is quite variable because a number of eggs in each batch usually were abnormal in their early development. Advanced and delayed hatchings are due to an early or late release of the hatching enzyme respectively. However, 50% of the embryos that follow the normal pattern of development hatched by about 108:00 A.F.

<u>Stage 40: yolk absorption at 140:00 A.F. (Fig. 40)</u> The ventral margin of the yolk sac, which has diminished greatly in size, lies on a straight line drawn along the ventral margin of the prolarva. The ventral caudal vein has completely atrophied; the central caudal vein now collects all arterial flow from the tail. The prolarva measures 5.24 mm in total length. There are 12 dermal fin rays in the caudal fin. Xanthophores are numerous along the dorso-lateral surface from head to tail but they are scarce laterally on the head, trunk and tail. A few snowflake-like guanophores are present on the dorso-lateral surface region extending from head to tail. Laterally the guanophores are small and few in number.

Development of pupfish embryos in sea water is largely the same as in fresh water. The time to reach the various stages of development in sea water (35 $^{0}/_{00}$) appears to be essentially the same as in fresh water including the hatching stage which occurs at 108:00 A.F. in both environments (Sweet and Kinne, personal communication). However, the temporal-morphological relationships of development in both environments may not be exactly the same.

SUMMARY OF STAGES

1 Stage 1, Unfertilized ovum showing numerous cortical alveoli, a central oil globule and a micropyle (arrow).

2 Stage 2, One-celled ovum. Small highly refractive oil globules larger than the cortical alveoli (which have disappeared) are scattered over the yolk.

3 Stage 3, Two-celled ovum; chorion not shown; 0.55 A.F. (after fertilization).

4 Stage 4, Four-celled ovum; 1:25 A.F.

5 Stage 5, Eight-celled ovum; 1:50 A.F.

6 Stage 6, Sixteen-celled ovum; 2:10 A.F.

7 Stage 7, Thirty-celled ovum. Two layers of cells are present; 2:30 A.F.

8 Stage 8, Early high blastula; 3:30 A.F.

9 Stage 9, Late high blastula. Aggregation of highly refractive oil globules at vegetal pole; 4:00 A.F.

10 Stage 10, Intermediate blastula. Blastoderm is not raised high off the yolk; 5:00 A.F.

11 Stage 11, Flat blastula. Blastula is not raised off the yolk; 6:00 A.F.

12 Stage 12, Early gastrula; top view of blastoderm. Central portion of blastoderm is thin whereas the periphery is thickened to form germ ring. Note the slight thickening of the germ ring to form the embryonic shield (arrow); oil globules remain central and below germ ring; 9:00 A.F.

13 Stage 13, Epiboly covering 1/3 yolk. Note periblast nuclei just above central oil globule and embryonic shield (arrow) just above the periblast; 11:00 A.F.

14 Stage 14, Middle gastrula. No periblast is visible; 12:00 A.F.

15 Stage 15, Epiboly covering 2/3 yolk. Embryonic shield (arrow) has thickened; 14:00 A.F.

16 Stage 16, Late gastrula and formation of optic anlagen; central oil droplet at top left corner; 15:00 A.F.

17 Stage 17, Formation of 2 pairs of somites. Note large blastopore at vegetal pole; 18:00 A.F.

18 Stage 18, Formation of auditory placodes and closure of the blastopore. Auditory placodes (arrow) are clearly visible; 19:00 A.F.

19 Stage 19, Formation of optic placodes. Optocoel is formed in optic anlagen; 20:00 A.F.

20 Stage 20, Early pigmentation; 21:00 A.F.

21 Stage 21, Formation of optic cups; 22:00 A.F.

22 Stage 22, Formation of auditory vesicles. The auditory placodes are vesicular (arrow); chorion to the left and yolk on the right; 24:00 A.F.

23 Stage 23, Expansion of optic lobes. The mesocoele (arrow) is quite prominent; oil globule is near left optic lobe; 26:00 A.F.

24 Stage 24, Initiation of heart beat. The heart (arrow) begins to beat without blood cell circulation; 28:00 A.F.

25 Stage 25, Increase in heart rate; 30:00 A.F.

26 Stage 26, Oscillation of blood cells in heart; 32:00 A.F.

27 Stage 27, Initiation of blood circulation. A few blood cells flow through the anterior vitelline veins; note chromatophores on yolk; oil globule is shown as dotted structure below embryo; 34:00 A.F.

28 Stage 28, Pectoral fin bud formation; 38:00 A.F.

29 Stage 29, Aggregation of otoliths; 41:00 A.F.

30 Stage 30, Blood flow in dorsal intersegmental arterioles. Blood cells circulate in the dorsal intersegmental tissue between two somites; 43:00 A.F.

31 Stage 31, Early retinal pigmentation; 47:00 A.F.

32 Stage 32, Intermediate retinal pigmentation. Retinal cells are outlined by thin rims of black pigment; 50:00 A.F.

33 Stage 33, Late retinal pigmentation. Retinal cells are outlined by thick rims of black pigment; two chromatophores on yolk are shown; 54:00 A.F.

34 Stage 34, Establishment of retinal pigment pattern; 63:00 A.F.

35 Stage 35, Blood circulation in pectoral fins. One pectoral fin is shown; 68:00 A.F.

36 Stage 36, Development of dermal fin rays in the caudal fin. Guanophores (black cells) form a distinctive pattern on caudal fin; 72:00 A.F.

37 Stage 37, Mouth open; 78:00 A.F.

38 Stage 38, Formation of central caudal vein. Central caudal vein is fully formed (left arrow) below the caudal artery (right arrow); note neural and hemal arches of the caudal vertebrae; 92:00 A.F.

39 Stage 39, Hatching; 108:00 A.F.

40 Stage 40, Yolk absorption; the yolk sac does not appear as an extension of the larvae. Only the pectoral and caudal fins are formed; 140:00 A.F.

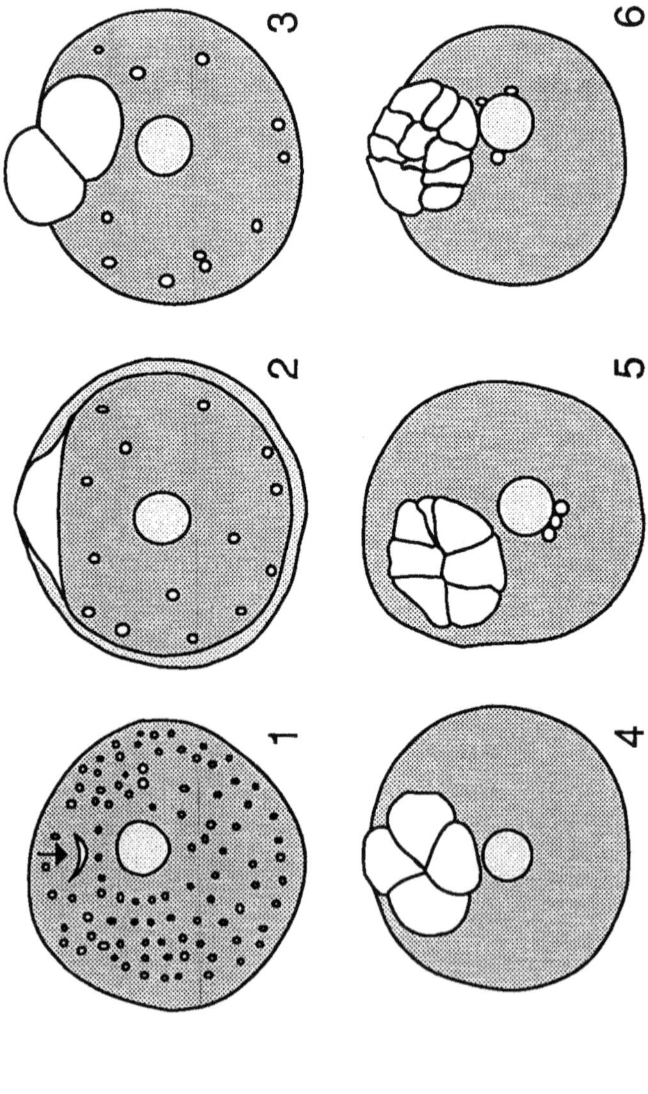

Summary of stages

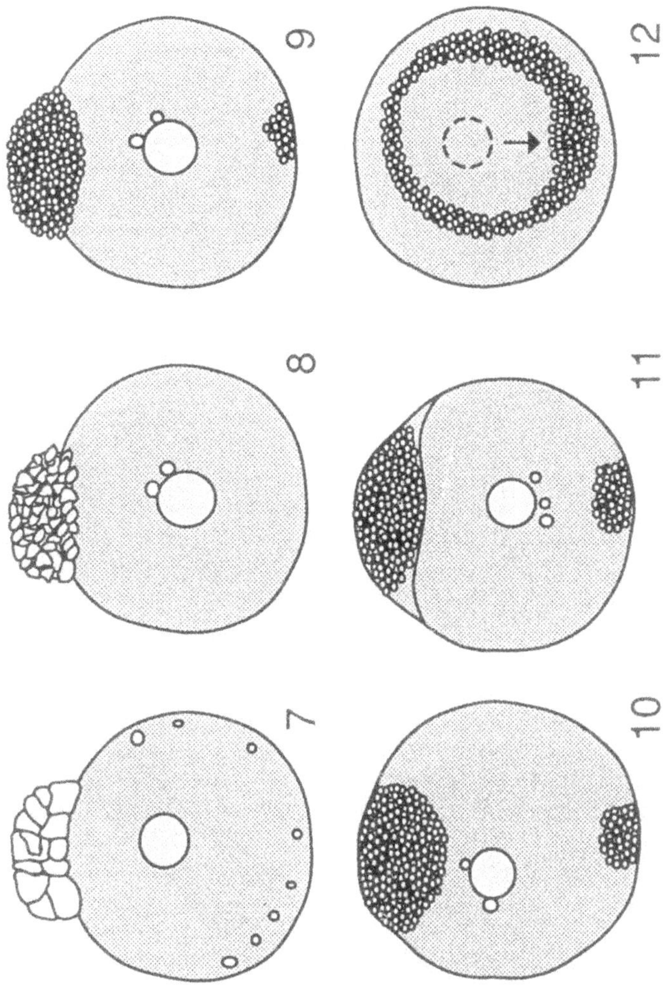

Summary of stages con't...

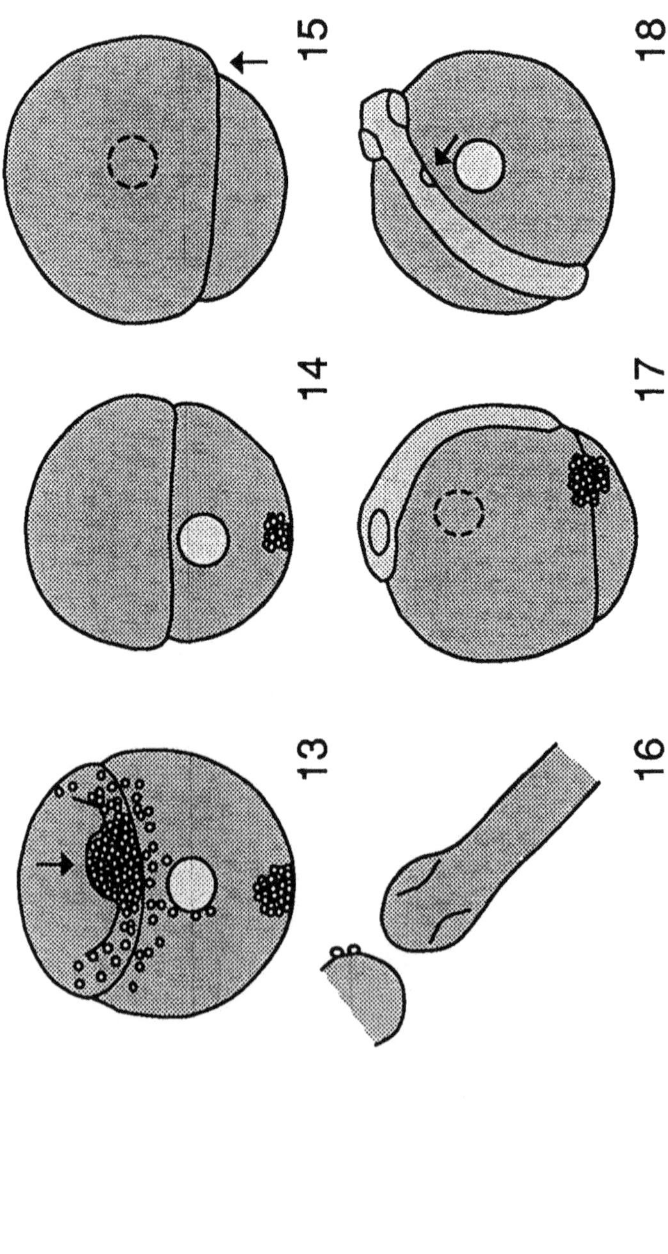

Summary of stages con't...

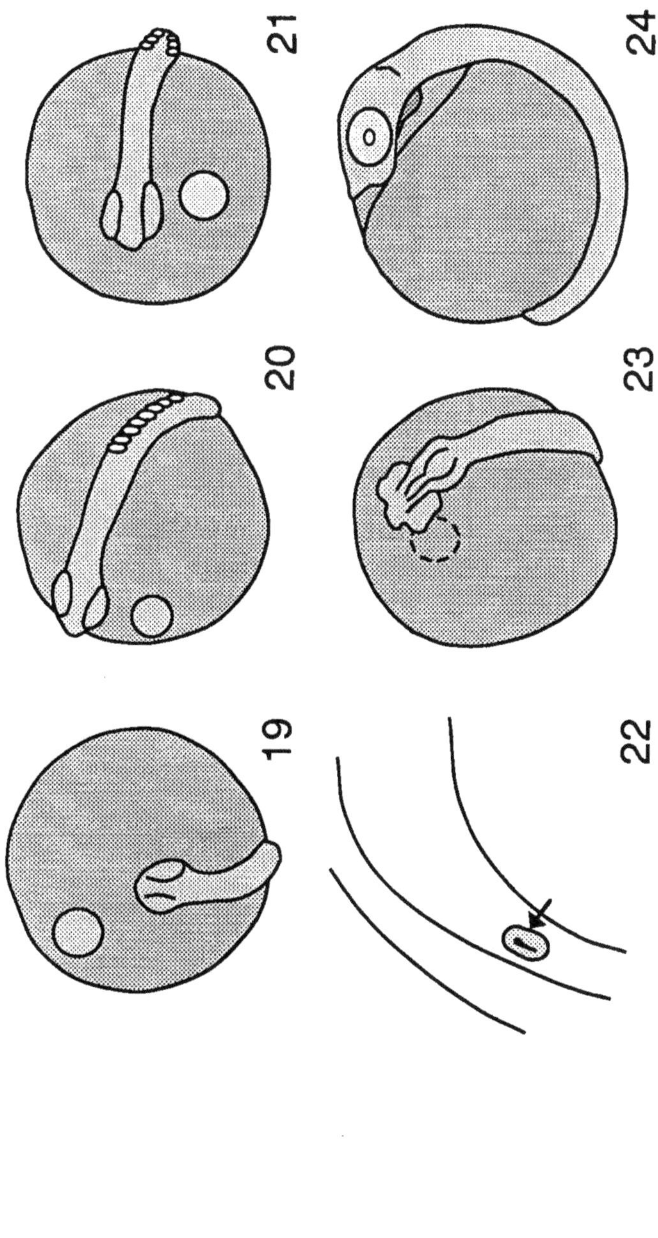

Summary of stages con't...

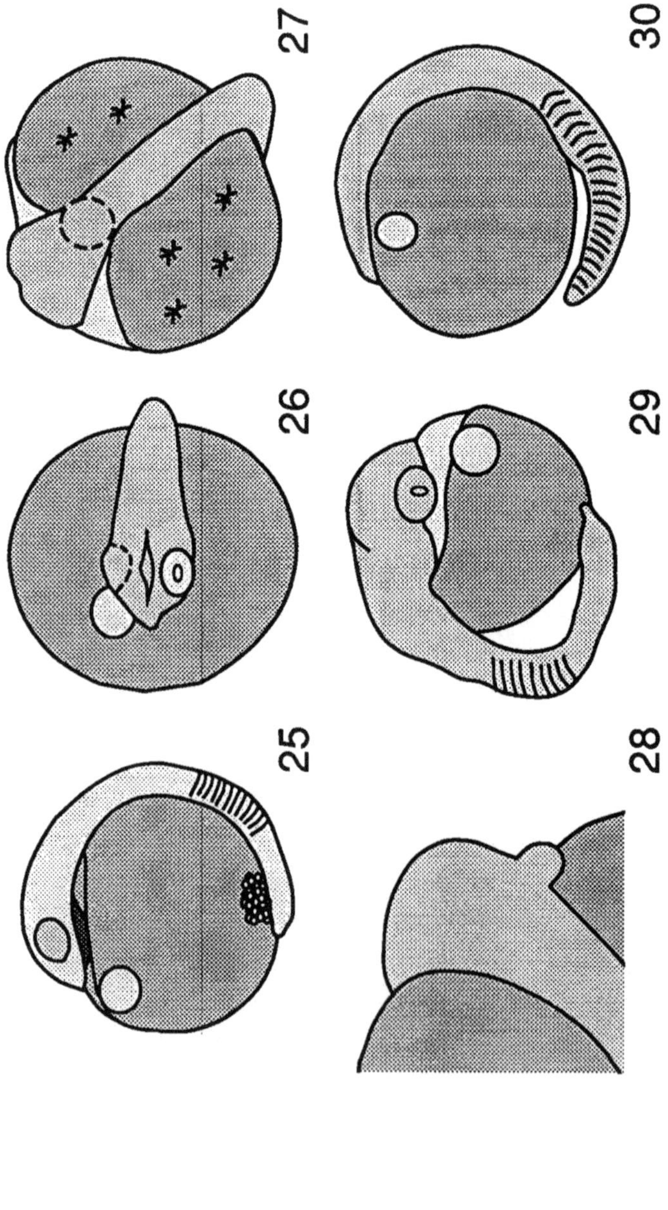

Summary of stages con't...

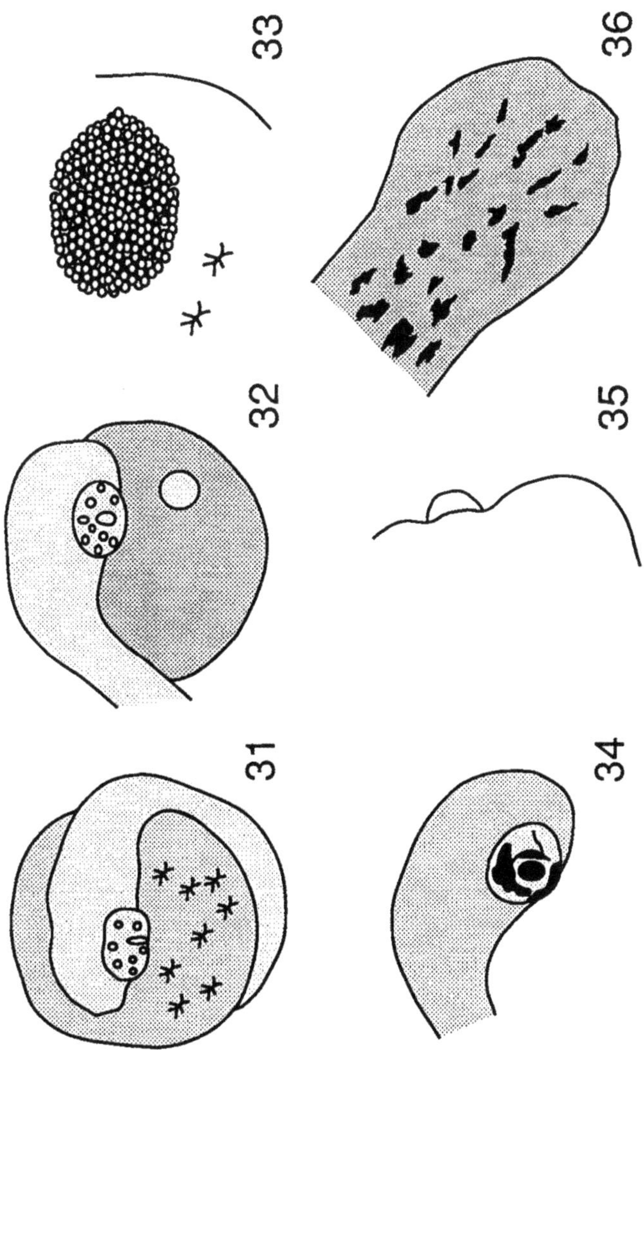

Summary of stages con't...

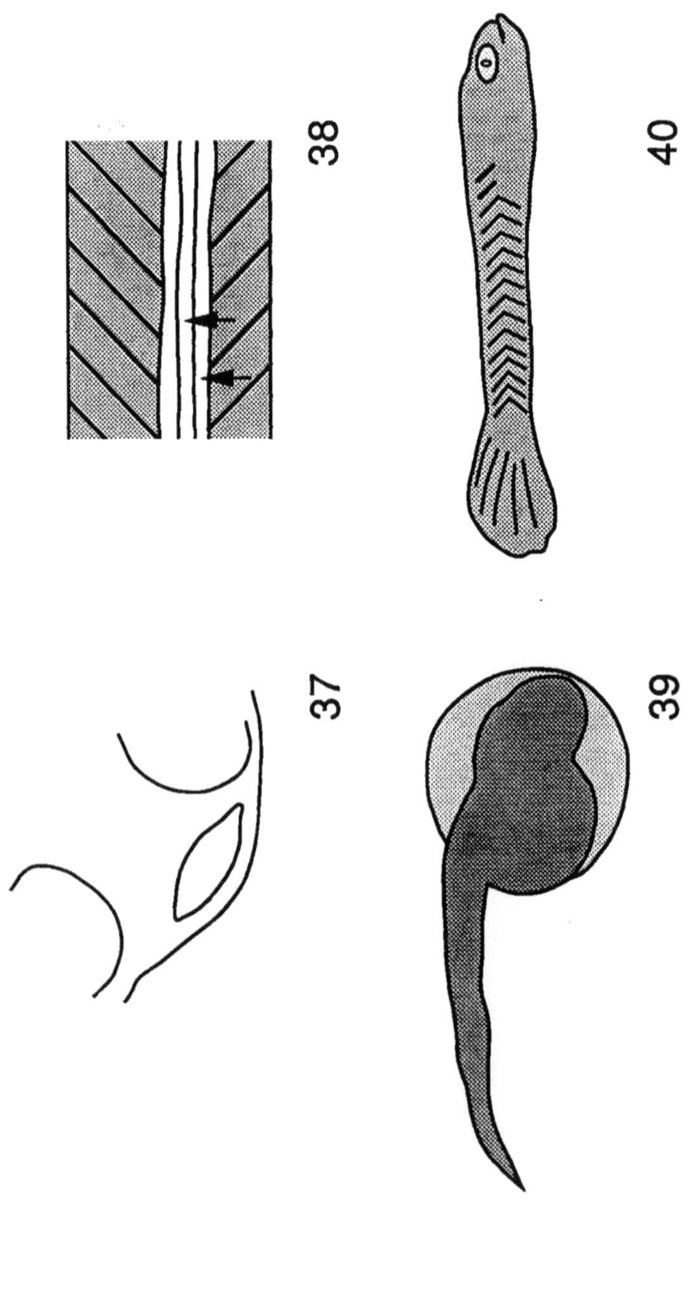

Summary of stages con't...

REFERENCES

Ando, S. (1962) Response of embryonic melanophores of the wild medaka (*Oryzias latipes*) to various stimuli. Embryologia 7: 169-178.

Baird, S.F. and Girard, C. (1853) Descriptions of new species of fishes collected by Mr. John Clark, on the U.S. and Mexican boundary survey, under Lt. Col. Jas. D. Graham. Proc. Acad. Nat. Sci. Phila. 6: 387-390.

Barlow, G.W. (1958a) Daily movements of desert pupfish, *Cyprinodon macularius*, in shore pools of the Salton Sea, California. Ecology 39: 580-587.

Barlow, G.W. (1958b) High salinity mortality of desert pupfish, *Cyprinodon macularius*. Copeia 1958: 231-232.

Barinaga, M. (1990) Zebrafish: swimming into the development mainstream. Science 250: 34-35.

Evans, J.W. (1961) Normal stages of the early development of the flying fish, *Hirundichthys affinis* Gunther. Bull. Mar. Sci. Gulf Caribbean 11: 483-502.

Garside, E.T. (1959) Some effects of oxygen in relation to temperature on the development of lake trout embryos. Can. J. Zool.37: 689-698.

Goodrich, H.B. (1927) A study of the development of Mendelian characters in *Oryzias latipes*. J. Exp. Zool. 49: 261-280.

Hayes, F.R. and Pelluet, D. (1945) The effect of temperature on growth and efficiency of yolk conversion on the salmon embryo. Can. J. Res. D 23: 7-15.

Hayes, F.R., Pelluet, D. and Gorham, E. (1953) Some effects of temperature on the embryonic development of the salmon. Can. J. Zool. 31: 42-51.

Hisaoka, K.K. and Battle, H.I. (1958) The normal developmental stages of the zebrafish, *Brachydanio rerio*. J. Morph. 102: 311-328.

Hisaoka, K.K. and Firlit, C.F. (1962) The embryology of the blue gourami, *Trichogaster trichopterus* (Pallas). J. Morph. 111: 239-253.

Hishida, T., Tomita, H. and Yamamoto, T. (1961) Melanin formation in colour varieties of the medaka (*Oryzias latipes*). Embryologia 5: 335-346.

Ho, R.K. and Kane, D.A. (1990) Cell-autonomous action of zebrafish spt-1 mutation in specific mesodermal precursors. Nature 348: 728-730.

Kaighn, M.E. (1962). A biochemical study of the hatching process in *Fundulus*, Ph.D. thesis. Massachusetts Institute of Technology.

Kaighn, M.E. (1964) A biochemical study of the hatching process in *Fundulus heteroclitus*. Dev. Biol. 9: 56-80.

Kinne, O and Kinne, E.M. (1962) Rates of development in embryos of a cyprinodont fish to different temperature-salinity-oxygen combinations. Can. J. Zool. 40: 231-253.

Mahon, H.F. and Hoar, W.S. (1956). The early development of the chum salmon, *Oncorhynchus keta* . J. Morph. 98: 1-47.

Maneckjee, A., Weisbart, M. and Idler, D.R. (1989) The presence of 17α20ß-dihydroxy-4-pregnen-3-one receptor activity in the ovary of the brook trout, *Salvelinus fontinalis*, during terminal stages of oocyte maturation. Fish Physiol. Biochem. 6:19-38.

Maneckjee, A., Idler, D.R. and Weisbart, M. (1991) Demonstration of putative membrane and cytosol steroid receptors for 17α,20ß-dihydroxy-4-pregnen-3-one in brook trout, *Salvelinus fontinalis*, oocytes by photoaffinity labelling using synthetic progestin 17,21-dimethyl-19-nor-pregn-4,9-diene-3,20-dione (R5020). Fish Physiol. Biochem. 9: 123-135.

Matsui, K. (1949) Illustration of the normal course of development in the fish, *Oryzias latipes*. Japanese J. Exp. Morph.5: 33-42.

Miller, R.R. (1943) The status of *Cyprinodon macularius* and *Cyprinodon nevadensis*, two desert fishes of Western North America. Occ. Papers Mus. Zool. Univ. Michigan No. 473: 1-25.

Miller, R.R. (1948) The Cyprinodont fishes of the Death valley system of eastern California and southwestern Nevada. Misc. Publ. Mus. Zool. Univ. Michigan No. 68: 1-155.

Nagahama, Y. (1983) The Functional Morphology of Teleost Gonads, In: Fish Physiology. Academic Press, New York, Vol IX A, pp. 223-275.

Nagahama, Y. (1987) Endocrine control of oocyte maturation, In: Hormones and Reproduction in Fishes, Amphibians, and Reptiles. D.O. Norris and R.E. Jones (eds.) Plenum Press, N.Y., pp. 171-202.

Nelson, O.E. (1953) Comparative Embryology of the Vertebrates. The Blackiston Company Inc. Toronto.

Nikol'skii, G.V. (1961) Special Ichthyology, 2nd edition Israel Program for Scientific Translations.

Oppenheimer, J.M. (1937) The normal stages of *Fundulus heteroclitus*. Anat. Rec. 68: 1-15.

Pelluet, D. (1944). Criteria for the recognition of developmental stages in the salmon (*Salmo salar*). J. Morph. 74: 395-407 .

Sakai, T.Y. (1961) Method for removal of chorion and fertilization of the naked egg in *Oryzias latipes*. Embryologia 5: 357-368.

Scott, A.P. and Canario, A.V.M. (1987) Status of oocyte maturation-inducing steroids in teleosts. In: Proceedings of the Third International Symposium on the Reproductive

Physiology of Fish. D.R. Idler, L.W. Crim and J.M. Walsh (eds.) Memorial University of Newfoundland, St. John's, pp. 224-234.

Solberg, A.N. (1938). The susceptibility of *Fundulus heteroclitus* embryos to X-radiation. J. Exp. Zool. 78: 441-469 .

Sumner, F.B. (1900) Kupffer's vesicle and its relation to gastrulation and concrescence. Mem. N.Y. Acad. Sci. 12.

Sweet, J.G. and Kinne, O., personal communication

Tavolga, W.N. and Rugh, R. (1947) Development of the platyfish, *Platypoecilus maculatus*. Zoologica 32: 1-15.

Trant, J.M., Thomas, P. and Shackleton, C.H.L. (1986) Identification of 17α,20ß21-trihydroxy-4-pregnen-3-one as the major ovarian steroid produced by the teleost *Micropogonias undulatus* during final oocyte maturation. Steroids 47: 89-99.

Trant, J.M. and Thomas, P. (1989) Changes in ovarian steroidogenesis in vitro associated with final maturation of Atlantic croaker oocytes. Gen. Comp. Endocrinol. 75: 405-412.

Tyler, C.R., Sumpter, J.P. and Bromage, N.R. (1988) In vivo ovarian uptake and processing of vitellogenin in the rainbow trout, *Salmo gairdneri*. J. Exp. Zool.246: 171-179.

Vernidub, M.F. (1956) Modifications in morphology and physiology of embryonal development in Ludoga whitefish (*Coregonus lavaretus ludoga Poljakow*). Tr. Karel'sk. Fil. Akad. Nauk S.S.S.R. 1956: 103-118.

Weisbart, M., Maneckjee, A. and Idler, D.R. (1990) Evidence of a membrane receptor for 17α20ß-dihydroxy-4-pregnen-3-one (DHP) in oocytes of salmonids. Amer. Zool. 30:(4), 52A.

Wilson, H.V. (1889) The embryology of the sea bass *Serranus atrarius*. U.S. Fish and Wildlife Serv. Fish. Bull.9: 209-223.

Yamamoto, M. (1963) Electron microscopy of fish development II. Oocyte-follicle cell relationship and formation of chorion in *Oryzias latipes*. J. Fac. Sci. Univ. Tokyo Sec.IV 10: 123-132.

Yamamoto, M. (1964) Electron microscopy of fish development III. Changes in the ultrastructure of the nucleus and cytoplasm of the oocyte during its development in *Oryzias latipes*. J. Fac. Sci. Univ. Tokyo Sec.IV 10: 335-344.

Yamamoto, K., Oota, I., Takano, K. and Ishikawa, T. (1965) Studies on the maturing process of the rainbow trout, *Salmo gairdneri irideus*. 1. Maturation of the ovary of a one-year old fish. Bull. Jpn. Soc. Sci. Fish.31: 123-132.

Young, G., Adachi, S. and Nagahama, Y. (1986) Role of ovarian thecal and granulosa layers in gonadotropin-induced synthesis of a salmonid maturation-inducing substance (17α,20ß-dihydroxy-4-pregnen-3-one). Develop. Biol.118: 1-8.

Zotin, A.I. (1958) The mechanism of hardening of the salmonid egg membrane after fertilization or spontaneous activation. J. Embryol. Exp. Morph. 6: 546-568.

CHAPTER 2

Gene Transfer and Expression in Medaka Embryos

Kenjiro Ozato, Koji Inoue, and Yuko Wakamatsu

ABSTRACT

The medaka, *Oryzias latipes* is a small fresh water teleost well suited as a model fish for transgenic studies. In this communication, we describe microinjection and electroporation techniques for gene transfer and several methods for the transgene's detection and analyses.

INTRODUCTION

The medaka (*Oryzias latipes*) is a small, egglaying freshwater teleost which is widely used as a laboratory animal (Yamamoto, 1975). The oogenesis, fertilization, and embryonic development of this species have been extensively studied. Table 1 summarizes biological characteristics of the medaka in relation to transgenic research. The 3 cm long medaka falls into the smallest group of vertebrates known. The generation time is short, 2 to 3 months, comparable with that of zebrafish and mice (see Vielkind, 1991, Chapter 5, this monograph). Spawning is daily and year-round under artificial conditions, and the timing of spawning can be controlled by the light cycle during a 24h period. In addition, the transparency of the eggs is a distinct advantage for embryological manipulations. Several foreign genes, including chicken δ–crystallin gene (Ozato et al., 1986; Inoue et al., 1989), bacterial CAT gene (Chong and Vielkind, 1989), rainbow trout growth hormone gene cDNA (Inoue et al., 1990a,b) and luciferase gene (Tamiya et al., 1990) have been introduced into the medaka. Here, we describe gene transfer into the medaka and the analysis of transgenes and gene products.

Table 1: Biological characteristics of the medaka

Strain	Orange-red-colored mutant Wild type
Inbred strain	Present
Life span	2 years
Sex maturation	3 months
Total length	3 cm
Spawning	Year-round
Spawning cycle	24h
Days for hatching	10 days
Diameter of eggs	1 mm
Transparency of eggs	Transparent
Chorion	Hard
Diameter of oocytes	1 mm
Diameter of germinal vesicles	100 μm
Culture of oocytes	Possible
Chromosome number	48
Genome size	20% of mammals

MEDAKA AS AN EXPERIMENTAL ANIMAL

Breeding of the Medaka

The cultivated orange-red-colored strain of the medaka is bred under controlled lighting (14h light and 10h darkness at 26°C). The medaka spawns at the beginning of the light period under these conditions.

Females and Males

Females and males are identified by the morphology of the dorsal and anal fins. The male anal fin is white and shaped as a parallelogram and the female one is yellow and gradually narrows towards the tail. The male dorsal fin is also slightly irregular in shape in comparison with that of the female. (Fig. 1).

Isolation of Females

Female medaka known to produce large numbers of eggs should be isolated on the day prior to experimentation. Spawning occurs just after the onset of the light cycle and the eggs remain attached to the abdomen. The normal ovulation cycle is 24h, therefore eggs can be collected from the same female on successive days (see also Vielkind, 1992, Chapter 5, this monograph).

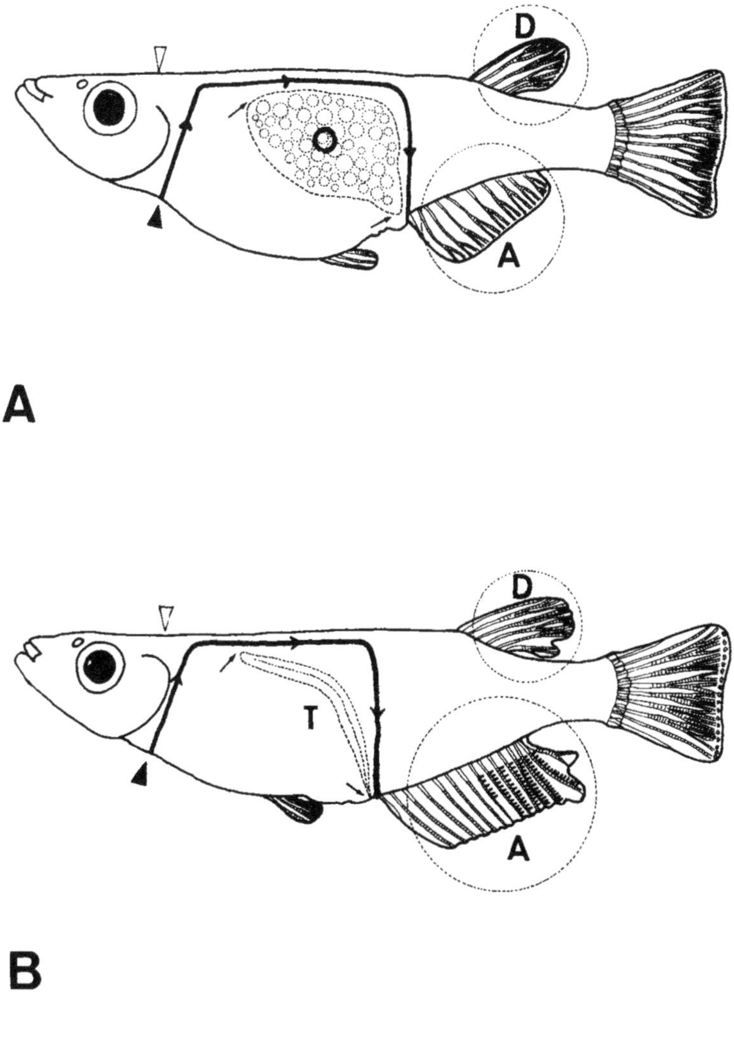

Fig. 1. Female (A) and male (B) of the medaka A, anal fin; D, dorsal fin; 0, ovary, T, testis

GENE TRANSFER BY MICROINJECTION

Small Tools for Microinjection

Egg holder: The nucleus of oocytes and the cytoplasm of one cell-stage embryos are located to one side of the animal pole. Prior to microinjection the eggs are oriented such that the needle can directly enter the animal pole and penetrate into the nucleus or cytoplasm (Fig. 2). The glass apparatus used to position and hold the eggs is also illustrated in Figure 2.

Plastic needle: A plastic needle is used for handling the eggs in the dishes or egg holders. The needle is made by attaching a fine fishing line to a glass tube.

Glass dish with a plastic mesh: Once the oocyte nucleus has been injected the follicle layer surrounding the egg is mechanically removed using forceps. This is done in a small glass dish with a plastic mesh covering the bottom of the inside. This mesh helps to prevent the eggs from rolling around.

Injector system with a pressure monitor : The microinjection apparatus is illustrated in Fig. 3. Microinjection is carried out by applying pressure to the injector. The pressure being applied to the injection system is monitored using an air filled 25 µl syringe built into the injection circuit. Changes in the volume of air in the syringe reflect changes in the injection pressure. The remaining open space in the injection system is filled with water, not oil.

Microinjection into the Oocyte Nucleus

Nine hours before ovulation preovulatory oocytes are at the prophase stage of the second maturation division. At this stage they are 900 to 1000 µm in diameter and located in the transparent cytoplasm at the animal pole (Fig. 4). In order to obtain these preovulatory oocytes for microinjection in early morning, the onset of the daily light period is set for 16:00 h.

Collection of oocytes

1. Prepare 1000ml Earle's 199 medium supplemented with 1.5g $NaHCO_3$, $5x10^4$ units penicillin, 30mg streptomycin, and 20g BSA (Sigma Fraction V) and sterilize by filtration.
2. Kill female adult fish by cutting through the back of the head 9-10h before ovulation (white arrowhead in Fig 1A).
3. Cut and open the abdominal cavity (along the black arrow line in Fig 1A), take out the ovary and put it into a dish filled with the modified Earle's medium.

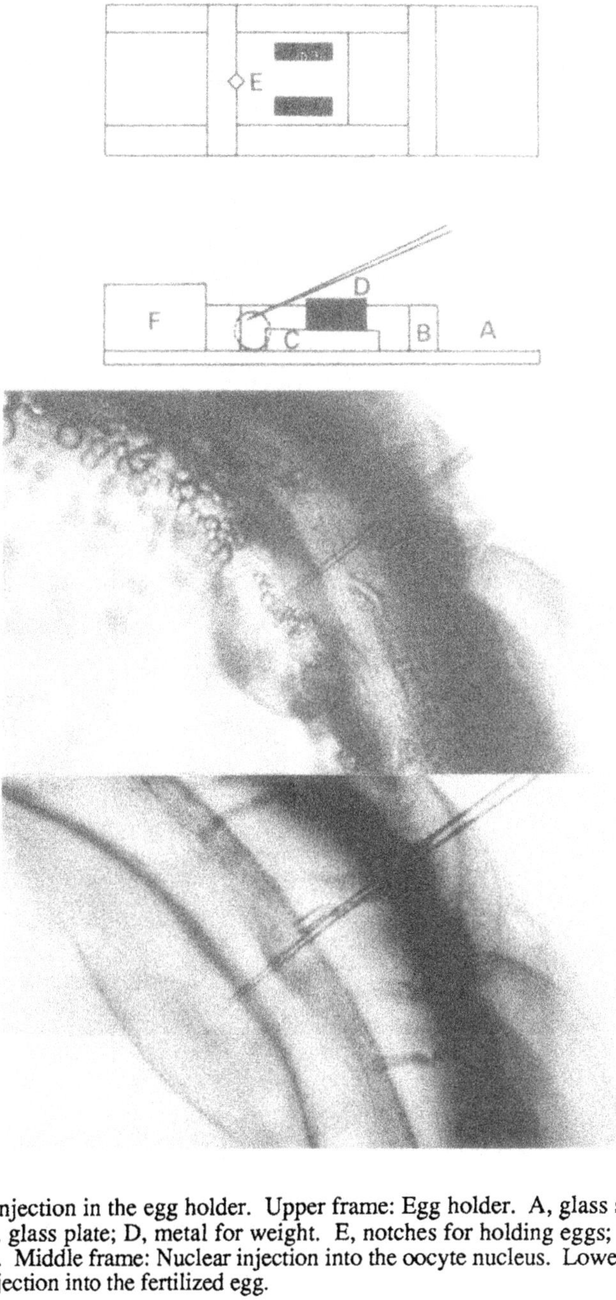

Fig. 2. Microinjection in the egg holder. Upper frame: Egg holder. A, glass slide; B, glass frame; C, glass plate; D, metal for weight. E, notches for holding eggs; F, glass rod for holding. Middle frame: Nuclear injection into the oocyte nucleus. Lower frame: Cytoplasmic injection into the fertilized egg.

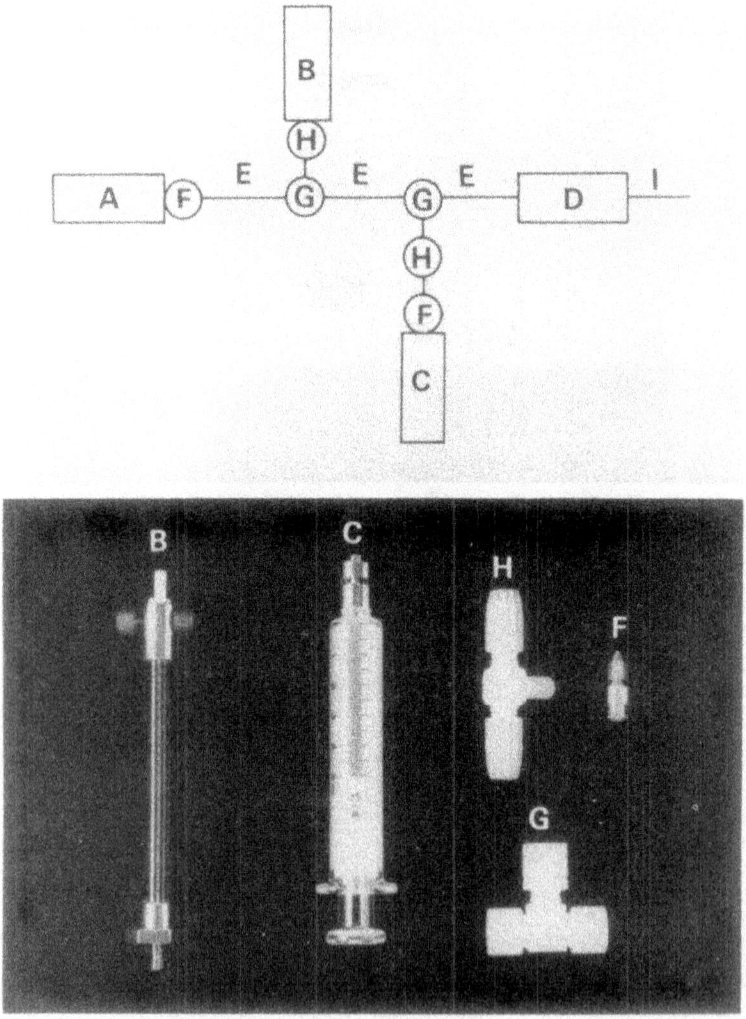

Fig. 3. Injector system. A, Injector; B, Pressure monitor (A 25 μl syringe with an opening or shunting cock at the tip and a glass tube projection at the base, (Precision Sampling, Baton Rouge); C, glass syringe for supplying water; D, needle holder; E, polyethylene tube (2mm outer diameter); F, connector between the injector or syringe and polyethylene tube (Narishige, Tokyo); G, T type miniature connector (used for chromatography); H, I type miniature connector with valve; I, glass needle.

4. Tear the ovary with forceps and isolate intrafollicular oocytes without spoiling the follicle layer.
5. Select oocytes which are transparent and largest in size. Remove tissue debris from the surface of oocytes (Fig. 4).

Microinjection

1. Place oocytes in the egg holder filled with the incubation medium and fix them under the dissecting microscope (Fig. 2).
2. Put the egg holder on the microscope stage.
3. Microinjection is carried out at 10 x 10 magnification using a steriomicroscope. Microneedles have tips sharpened to 3-4 μm inner diameter.
4. Insert needle into the nucleus and inject the DNA solution. The volume injected is estimated by the degree of swelling of the nucleus. At this point, the injection volume is 20-30 pl. Injection of a larger volume results in degeneration of the embryos before the blastula stage. The injection pressure under these conditions, as estimated by the pressure monitor, is 1.2 to 1.4 atmospheres.
5. Prior to injection, the DNA solution is centrifuged to remove any particulate matter. The DNA solution consists of tris-EDTA, phosphate buffered saline or water containing 10 μg/ml DNA. Higher concentrations of DNA may be toxic or result in abnormal embryonic development.
6. As the oocytes mature in the incubation medium, the nucleus gradually flattens and the nuclear membrane breaks down and hampers microinjection. Therefore the time for microinjection is limited to about 2 h after oocyte collection. During this time 60-100 oocytes can be injected.

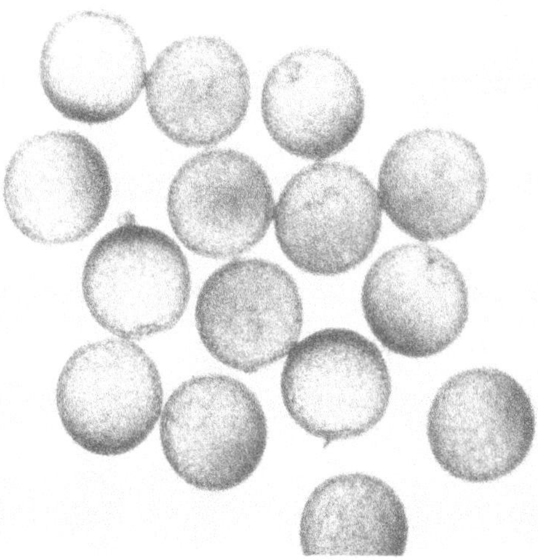

Fig. 4. Medaka oocytes.

Oocyte culture and artificial insemination

1. Put the injected oocytes into a dish incubation medium and keep them at 26°C for about 8h until the anticipated time of ovulation (16:00, see Fig. 5).

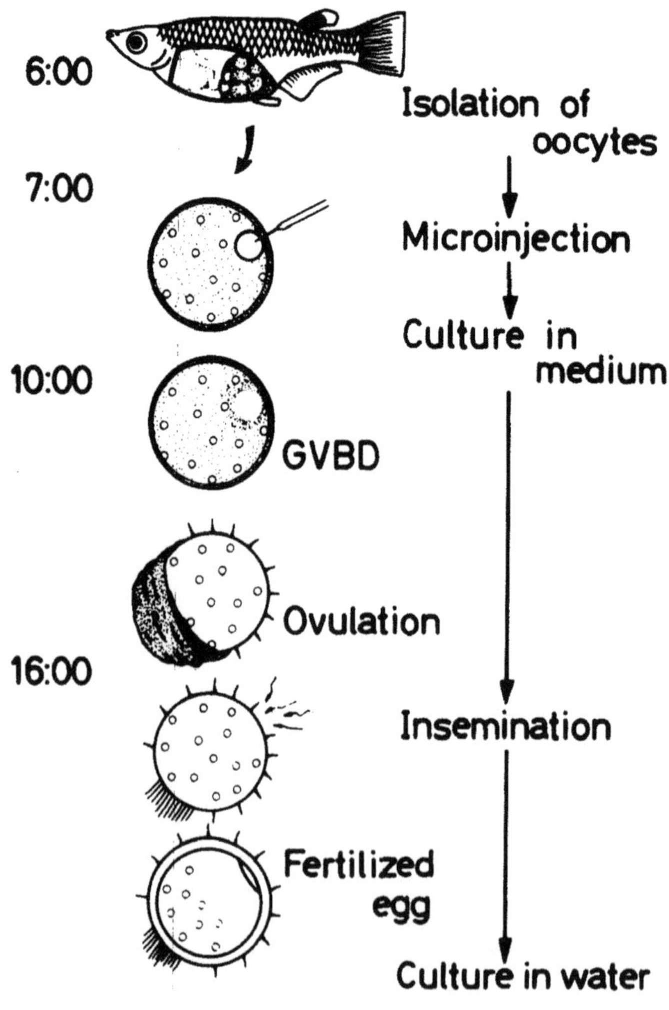

Fig. 5. The time schedule for microinjection into the medaka oocyte nucleus. GVBD, germinal vesicle breakdown.

2. During maturation of the cultured oocytes some of them automatically shed their follicle layer. The remainder have to have their follicle removed mechanically. To do this the oocytes are placed in a small glass dish containing a layer of plastic mesh. The follicle layer is peeled from the oocyte, starting at the vegital pole where thick bundles of filaments are attached (Fig. 6).
3. Transfer the follicle-free oocytes to a hollow glass slide containing a small volume of incubation medium.
4. Kill adult male fish and remove testes using procedures similar to those described for removing the ovaries from females (Fig. 1B). Prepare sperm suspension by mincing the testes in a small volume of incubation media using forceps.
5. Pour the sperm suspension into the hollow glass slide containing the unfertilized oocytes and mix gently. Approximately 60% of the injected eggs are successfully fertilized using these procedures.
6. After an hour of incubation, cut the attaching filament from the fertilized eggs to avoid their aggregation (Fig. 7).
7. Wash the fertilized eggs three times with distilled water.
8. Incubate each egg separately in distilled water in 96-well microplates at 26°C. Approximately 50% of the injected eggs survive under these experimental conditions. Approximately 50% of the survivors carry the injected gene.

Microinjection into the Fertilized Egg Cytoplasm

Microinjection into the cytoplasm of fertilized eggs is the more general method of producing transgenic fish (Ozato et al., 1989). This procedure is easier than nuclear injection because it does not require oocyte culture, removal of the follicle or artificial insemination. In addition the fertilized eggs can be readily obtained without killing the fish. The one problem associated with cytoplasmic injection of fertilized eggs is the short time between fertilization and first cleavage: 60 minutes at 26°C. Thus the period for microinjection into the one celled embryo is only 30-40 minutes. This limits the number of eggs that can be injected from a single fish. In order to increase the number of eggs that can be used in an experiment the time of mating between the pair of medaka is staggered (Inoue et al., 1990a, b). Alternatively the time to first cleavage can be reduced considerably by incubating the fertilized eggs at low temperatures (Chong and Vielkind, 1989).

Procedures:

1. Under controlled photoperiod (14h light and 10h darkness), females spawn at the beginning of the light period.
2. Females and males are separated one day before the experiment and kept separately during the dark period.
3. Mate females and males at the beginning of the light period. Females usually spawn a few minutes after mating.
4. Take egg clusters from the abdomen of the females 20 min after mating. Cut off the attaching filaments with a fine scissors (Fig. 7).
5. Fix an egg in the egg holder (Fig. 2) and inject about 100-500pl of DNA solution at a concentration of 100 μg DNA/ml in PBS into the germ disk before the first

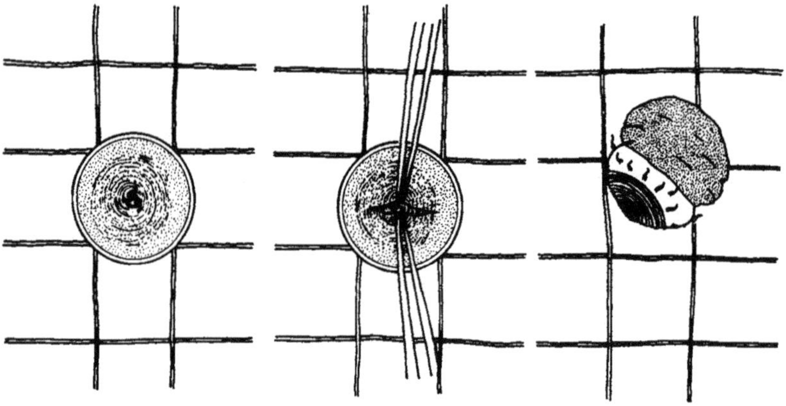

Fig. 6. Removal of the follicle layer.

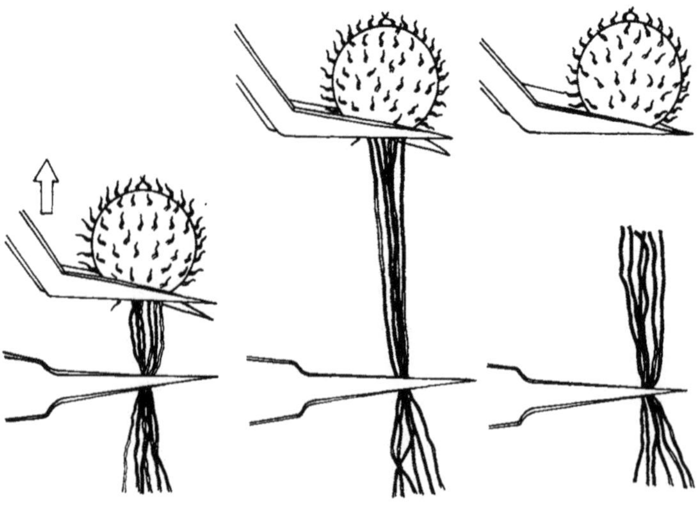

Fig. 7. Cutting the attaching filaments by micro-scissors.

cleavage (Fig. 2). The tip of microneedles must be sharpened, otherwise they cannot penetrate through the chorion and cell membrane.

6. Incubate injected eggs separately in 96-well microplates filled with distilled water.
7. By cytoplasmic injection, about 80% of injected embryos survive and about 50% of the surviving ones carry the foreign gene.

GENE TRANSFER BY ELECTROPORATION

Foreign gene transfer into medaka by electroporation has been demonstrated by Inoue et al., (1990a, b). This procedure is relatively easy for it requires no skilled techniques. Its one drawback at the present time is that the efficiency of gene transfer is lower than that obtained using nuclear or cytoplasmic injection methods.

The following procedures are used for introducing rainbow trout growth hormone cDNA into medaka.

1. Collect fertilized eggs as described in 1-4 of the previous section.
2. Put eggs between the electrodes of a transfection chamber (Shimazu FTC-03, 2mm electrode distance) filled with 800 μl of DNA solution (100μ1 g/ml PBS). About 120 eggs can be treated at a time.
3. Apply electric pulses (Shimazu GTE-1) using the following conditions: pulse height, 75OV/cm; pulse interval, ls; pulse length 50 μs; pulse number, 5.
4. Rinse treated eggs with distilled water several times and incubate separately at 26°C in 96-well plastic microplates filled with distilled water.
5. Using these experimental conditions, about 10% of embryos die immediately from damage to the yolk sac (Fig. 8A). Of the surviving embryos, 25% hatch at 8-20 days after fertilization. 4% of hatchlings are transgenic (Fig. 8B).

ANALYSIS OF TRANSGENES AND GENE PRODUCTS

Southern Blot Analysis

Standard procedures are used for DNA extraction and Southern blot analysis (Sambrook et al., 1989). Here, procedures for DNA extraction from 7-day-old medaka embryos are described.

1. Remove the chorion with fine forceps in PBS under a dissecting microscope. Transfer the embryos into microfuge tubes. Remove the saline with Pasteur pipettes.
2. Add 200 μl of lysis buffer containing 10mM Tris-HCl(pH7.5), l0mM EDTA, 150mM NaCl, 0.5% SDS and 0.4 mg/ml Proteinase K to each tube. Incubate at 55°C for 4h.
3. Extract the solution twice with an equal volume of buffer-saturated phenol and then twice with phenol/chloroform /isoamyl alcohol (25: 24: 1).
4. Precipitate the DNA with ethanol.
5. Recover the precipitate by centrifugation.

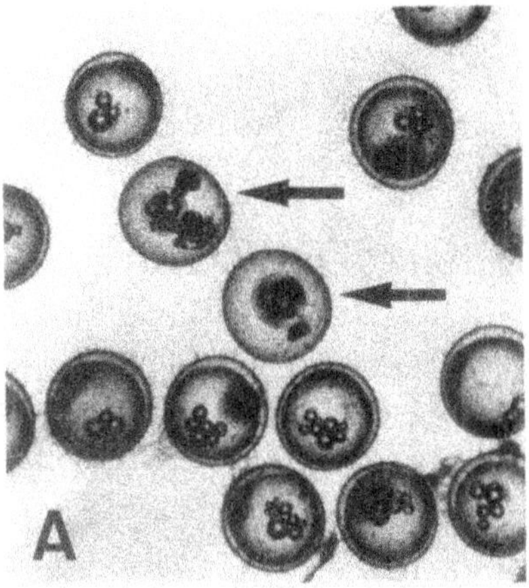

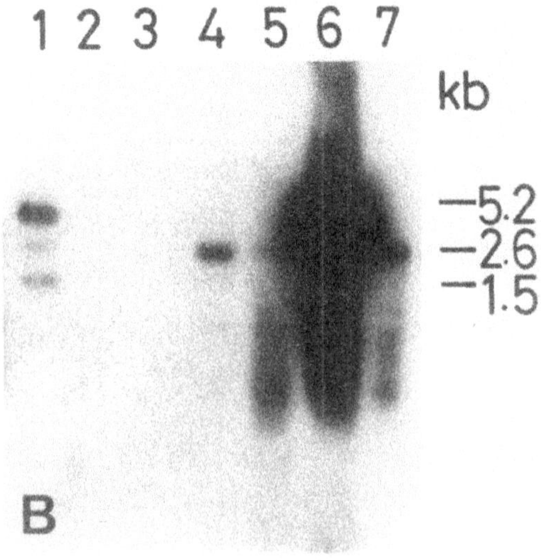

Fig. 8. Gene transfer by electroporation. Rainbow trout growth hormone cDNA was introduced into the medaka egg by electroporation (Inoue et al., 1990a,b). A: Early embryos 5h after electroporation. Several embryos are damaged by electric pulse (arrows). B: Southern blot analysis of DNA from a medaka fry which was treated with electric pulses at the one-cell stage. The correct size of the transgene is 2.6kb. The transgene is detected in lanes 1 and 4-7.

6. After drying, resuspend the pellet in 10mm Tris-HCl (pH8.0), 1mM EDTA. About 1.5 to 2 μg DNA is isolated from each 7-day-old embryo. This amount is enough for one Southern analysis.
7. An example of Southern blot analysis of DNA from 7-day-old medaka embryo is shown in Fig. 8.

DNA-DNA in situ Hybridization

This method is essential to examine the histological distribution of the transgenes (Takahashi et al., 1985). In addition, immunohistochemical staining techniques coupled with DNA-DNA in situ hybridization make it possible to study the cellular expression of the transgene in relation to its presence (Ozato et al., 1986).

The following methods were used to examine the tissue distribution and expression of the chicken δ-crystallin gene (Fig. 9A).

1. Remove the chorion with fine forceps in PBS under a dissecting microscope.
2. Fix dechorionated embryos with Carney's fixative or 4% paraformaldehyde in PBS at room temperature for 30min.
3. Embed embryos in paraffin and section at 5 μm thickness. Affix sections to glass slide.
4. For denaturation of DNA with alkali, treat deparaffinized sections with a solution of 0.5M NaOH and 1.5M NaCl at room temperature for 1.5min.
5. For neutralization, transfer the slides to 0.5M Tris-HCl (pH7.4) containing 1.5M NaCl.
6. Rinse the slides in 2 x SSC (1 x SSC is 0.15M NaCl, 0.015M sodium citrate, pH7.5).
7. Fix the sections again in 4% paraformaldehyde in PBS.
8. Treat the sections with prehybridization buffer of 6 x SSC containing 100 μg/ml denatured salmon DNA, 0.012% (w/v) heparin and 0.12% (w/v) each of Ficoll and polyvinylpyrrolidone at 68°C for 2h.
9. Label the probe with ^{3}H-dATP (60μCi/mmol), ^{3}H-dCTP (50 μ Ci/mmol) and ^{3}H-dTTP (90μ Ci/mmol) by nick translation to a specific activity of 2 x 10^8 dpm/ 4 μg DNA.
10. Treat the sections under cover slips with the hybridization buffer of 6 x SSC containing 200ng probe DNA/ml, 10% (w/v) dextran sulfate, 50 μg/ml denatured salmon DNA, 10mM EDTA, 10mM sodium phosphate, 0.012% (w/v) heparin and 0.12% (w/v) each of Ficoll and polyvinylpyrrolidone at 68°C for 30h.
11. Remove cover slips in 6 x SSC at room temperature.
12. Wash the slides with several changes of 2 x SSC at 50°C for 2h and dry.
13. Coat the slides with autoradiographic emulsion NTB-2 (Kodak) and expose them at 4°C for 4 weeks.
14. Stain the sections with hematoxylin.

Immunohistochemical Staining

Several methods are available for immunohistological detection of proteins produced by transgene expression. The peroxidase-antiperoxidase (PAP) method is popular. However the avidin-biotin-peroxidase complex method (ABC) can also be used (Hsu et al., 1981). Various kits are available for this method (Vector Laboratories, Burlingame).

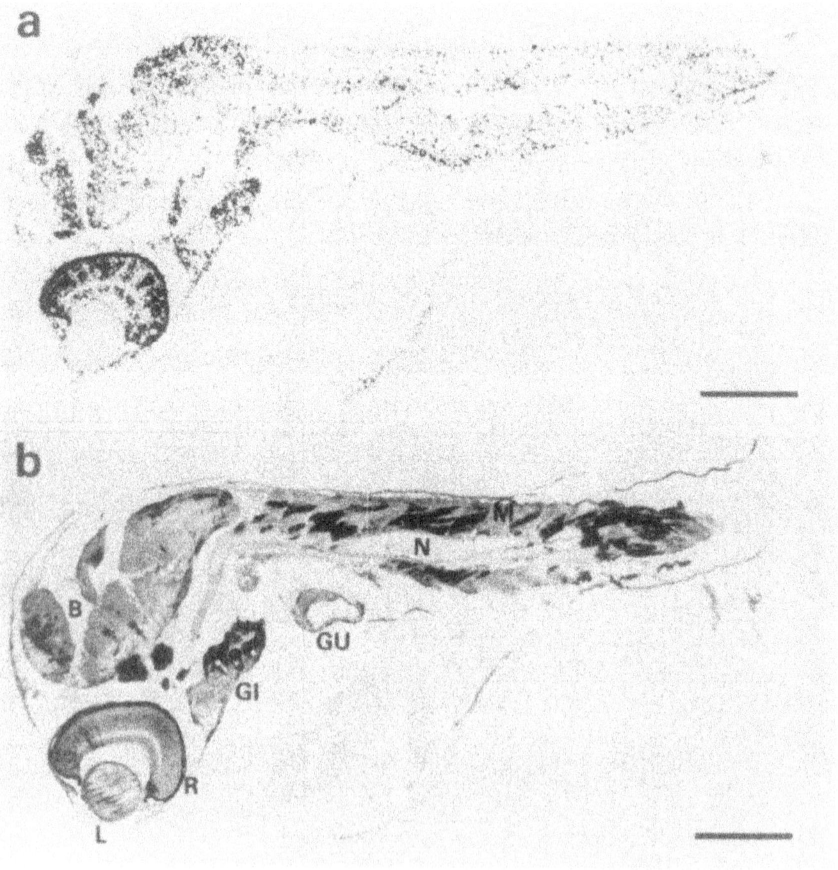

Fig. 9. Histological analysis of transgenes and gene products. a: The chicken δ-crystallin gene detected by DNA-DNA in situ hybridization in a 7-day-old medaka embryo. b: Chicken δ-crystallin polypeptides detected by immunohistochemical staining in the same embryo (Ozato et al., 1986). B, brain; GI, gill; GU, gut; L, lens; M, muscle; N, notochord; R, retina.

The following method was used to detect δ-crystallin in 7-day-old transgenic medaka embryos (Fig. 9B).

1. Remove the chorion with fine forceps in PBS under the dissecting microscope.
2. Fix dechorionated embryos with 4% paraformaldehyde in PBS.
3. Embed embryos in paraffin and section at 5 μm thickness. Affix sections to glass slide.
4. Dip deparaffinized slides in methanol containing 0.75% H_2O_2 for 30min to inhibit endogenous peroxidase activity.
5. Rinse the slides in the washing buffer containing 50 mM Tris-HCl and 1 mM EDTA (pH7.5) for 1h.
6. Treat sections with 5% normal goat serum in PBS for 30min to avoid nonspecific binding of the antibody. All procedures using sera and antibodies are carried out under cover slips in a moist chamber.
7. Treat the sections with the rabbit anti-δ-crystallin serum which is diluted to a 1/300 with PBS for lh.
8. Rinse the slides in the washing buffer for over lh.
9. Treat the sections with the goat anti-rabbit IgG serum (Cappel, Malvern) diluted with PBS to a 1/20 for lh.
10. Rinse the slides in the washing buffer for over lh.
11. Treat the sections with rabbit PAP (Cappel, Malvern) diluted with PBS to a 1/100 for lh.
12. Dip the slides in 0.1 M sodium acetate (pH5.2, adjusted with acetic acid).
13. Add substrate solution to each slide and react for 10 min or more. The substrate solution is prepared by mixing 4.75 ml of 0.1M sodium acetate (pH5.2), 0.25 ml of 0.4% 3-amino-9-ethylcarbazole in dimethylformamide and 5 μl of 30% H_2O_2 just before use.
14. Rinse the slides in the washing buffer for 5min.
15. After counter staining with hematoxylin, mount the slides with glycerol. Deposits of the pigment resulting from peroxidase reaction are readily soluble in organic solvents.

Immuno-blot Analysis

The following method was used to detect chicken δ-crystallin in 7-day-old medaka embryos (Fig. 10).

1. Remove the chorion in PBS with fine forceps under the the dissecting microscope. The yolk protein is removed by tearing the yolk sac to prevent non-specific binding. Transfer the embryos into microfuge tubes.
2. Add 10 μl of sample buffer containing 125 mM Tris-HCl (pH6.5), 2% sodium dodecylsulfate (SDS), 20%glycerol, 0.1% bromophenol blue, phenylmethyl-sulfonyl fluoride (PMSF) and 5% ß-mercaptoethanol. Heat in boiling water for 10min.
3. Electrophorese in a polyacrylamide gel slab.
4. Transfer the proteins onto a nitrocellulose sheet with electro-blotter.
5. To block the non-specific binding of the antibody, immerse the sheet in 0.5% skim milk (Difco) containing 50 mM Tris-HCl (pH8.0), 0.15M NaCl, 1 mM EDTA at 4°C for 16h.

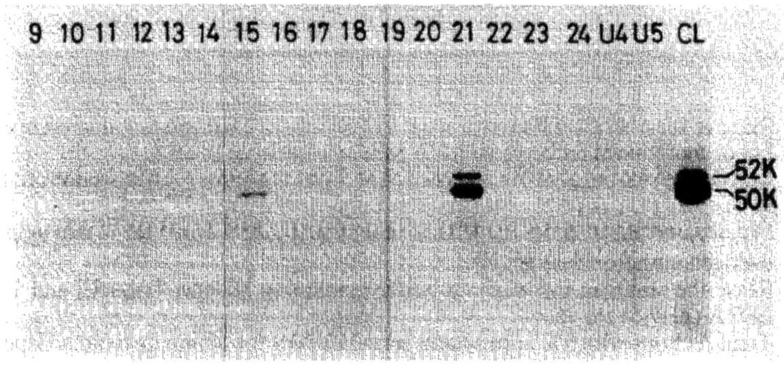

Fig. 10. Western blot analysis of transgene expression. Expression of the chicken δ-crystallin gene introduced into medaka embryos is detected in 7-day-old embryos. Lanes 9-24, injected embryos; U4-U5, non-injected embryos; CL, chicken embryo lens (Ozato et al., 1986).

6. Treat the sheet with the anti- δ-crystallin rabbit serum diluted to a 1/300 with the skim milk solution described in step 5 for lh.
7. Wash the sheet 2-3 times in a solution containing 0.02% Tween 20, 5OmM Tris-HCl (pH8.0), 0.15M NaCl, 1mM EDTA each for 15min.
8. Treat with ^{125}I-protein A diluted to 0.5 μ Ci/ml with the skim milk solution described in step 5 for lh.
9. Rinse the sheet several times in the Tween 20 solution described in step 7 each for 15min.
10. Dry the sheet and expose to X-ray film.

If the radioisotope is not available, use the peroxidase reaction and follow steps 5 to 14 of immunochemical staining procedures.

ACKNOWLEDGMENTS

We thank Dr. H. Kondoh for his critical reading of the manuscript. Thanks are also due to Ms. N. Kasai and Mr. S. Nagayama for drawing the figures. This article is written on the basis of the work supported in part by grants from the Ministry of Education, Science and Culture (No. 01619502) and the Fisheries Agency, Japan.

REFERENCES

Chong, S.S.C. and Vielkind, J. R. (1989) Expression and fate of CAT reporter gene microinjected into fertilized medaka (*Oryzias latipes*) eggs in the form of plasmid DNA, recombinant phage particles and its DNA. Theor. Appl. Genet. 78: 369-380.

Hsu, S-M., Raine, L. and Fanger, H. (1981) A comparative study of the peroxidase-antiperoxidase method and an avidin-biotin complex method for studying polypeptide hormones with radioimmunoassay antibodies. Am. J. Clin. Pathol. 75: 734-738.

Inoue, K., Ozato, K., Kondoh, H., Iwamatsu, T., Wakamatsu, Y., Fujita, T. and Okada, T. S. (1989) Stage-dependent expression of the chicken δ-crystallin gene in transgenic fish embryos. Cell. Differ.Dev. 27: 57-68.

Inoue, K., Asada, S., Kabeno, S., Yamashita S., Nagahisa, E. and Fujita, T. (1990a) Production of transgenic medaka as a model for genetic engineering of fish. In Current Topics in Marine Biotechnology. Miyachi, S., Karube, I. and Ishida, Y. (eds.) Fuji Technology Press, Tokyo. pp. 277-280.

Inoue, K., Yamashita, S., Hata, J., Kabeno, S., Asada, S., Nagahisa, E. and Fujita, T. (1990b) Electroporation as a new technique for producing transgenic fish. Cell Differ Dev. 29: 123-128.

Ozato, K., Kondoh, H., Inohara, H., Iwamatsu, T., Wakamatsu, Y. and Okada, T.S. (1986) Production of transgenic fish: introduction and expression of chicken δ-crystallin gene in medaka embryos. Cell Differ. Dev. 19: 237-244.

Ozato, K., Inoue, K. and Wakamatsu, Y. (1989) Transgenic fish: biological and technical problems. Zool. Sci. 6: 445-457.

Sambrook, J., Fritsch, E. F. and Maniatis, T. (1989) Molecular Cloning, A Laboratory Manual (2nd ed). Cold Spring Harbor Laboratory Press, Cold Spring Harbor, NY.

Takahashi, Y., Okada, T. S. and Kondoh, H. (1985) Histological detection of chicken δ-crystallin DNA sequences introduced into mouse teratocarcinoma cell lines. Dev. Growth Differ. 27: 607-613.

Tamiya, E., Sugiyama, T., Masaki, K., Hirose, A., Okoshi, T. and Karube, I. (1990) Spatial imaging of luciferase gene expression in transgenic fish. Nucleic Acids Res. 18: 1072.

Vielkind, J. R. (1992) Medaka and Zebrafish: Ideal as transient and stable transgenic systems. In Transgenic Fish. C. L. Hew and G. L. Fletcher (eds.) World Scientific Publishing Co. Singapore. pp. 72-91.

Yamamoto, T. (1975) Medaka (Killifish): Biology and Strains. Keigaku Publishing Company, Tokyo.

CHAPTER 3

Gene Transfer in Salmonids by Injection Through the Micropyle

Margaret A. Shears, Madonna J. King, Sally V. Goddard
and Garth L. Fletcher

ABSTRACT

Our technique and apparatus used to microinject gene constructs into salmonid eggs is described in detail. In brief the procedure is as follows. The injection apparatus consists of a glass needle 3-5μm in diameter driven by a micromanipulator. The eggs are fertilized in a salmon Ringer solution to delay hardening of the chorion. The DNA solution is injected into the germinal vesicle area by sliding the needle into the egg through the micropyle. The volume injected (2-3nl) is regulated using short bursts of N_2 using a "gene-pusher". Survival of the eggs to first feeding is usually high (approx. 80%) and genomic integration frequencies are approximately 3%. These injection procedures and results are discussed in comparison with the published results of others working with salmonids.

INTRODUCTION

It has been generally assumed that direct injection of gene constructs into the pronucleus or nucleus of the egg before the onset of cleavage is the most effective way to obtain incorporation of the new genetic material. The rational behind this is that integration into the genome prior to first cleavage will result in all of the organism's cells containing the same copy number of the foreign gene integrated into the same chromosomal sites. Although it is evident that this result can be achieved in mammals (Strojek and Wagner, 1988), available evidence from studies on fish are not as encouraging. The production of mosaics, indicating integration at or after the two cell stage, appears to be the rule rather than the exception.

One of the difficulties encountered with fish eggs lies with the inability of most investigators to visualize the nucleus or pronucleus using conventional light microscopy. In most, if not all cases, this is attributable to the opaqueness of the chorion and/or the cytoplasm. Rokkones et al., (1989) reported unsuccessful attempts to increase the visibility of the pronucleus in salmonid eggs by centrifugation or by vital staining using fluorescent dyes. Centrifugation has proved effective with pig eggs (Hammer et al., 1985). Yamaha et al., (1988) has successfully located the female pronucleus in dechorionated goldfish eggs using the fluorescent dye, Hoechst 33342. However no studies have been carried out to determine the effects of this dye or the ultraviolet light necessary for its fluorescence, on the viability of the eggs. Hammer et al., (1985) have used this technique on mammals and report that the procedure damages the ovum.

Since nuclear structures cannot usually be located, most investigators have opted for cytoplasmic injections following fertilization. Cytoplasmic injections, although not as effective as nuclear injections, have been successful in mice (Brinster et al., 1985). The target for cytoplasmic injections in fish eggs is the thin layer of ooplasm just under the chorion or the developing blastodisc (Fig. 1). If the needle penetrates the yolk, the eggs usually die.

Egg Structure

The large size of most teleost fish eggs is due primarily to their quantity of yolk (Fig. 1). Upon ovulation and release from the body cavity (spawning) the egg is

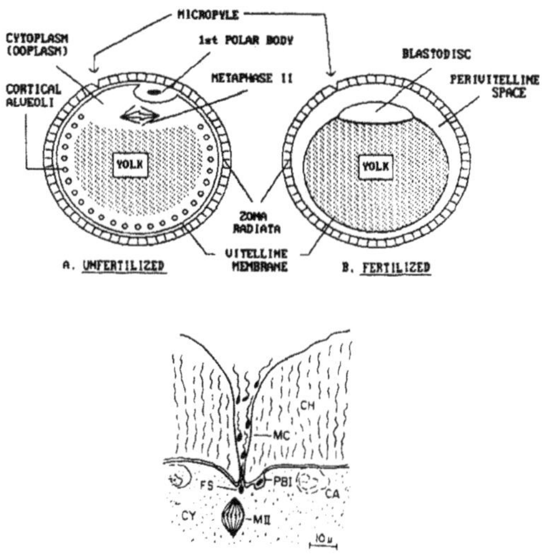

Fig. 1. Diagram of a salmonid egg. CH = Chorion, MC =micropilar canal, FS = Fertilizing sperm, PBI = first polar body, CA = cortical alveoli, MII = metaphase II, CY = cytoplasm.

encased in a thick proteinaceous outer membrane that serves to protect the egg. This membrane is known as the chorion or zona radiata (30-60 μ thick, salmonids). This membrane is impermeable to sperm. All teleost fish eggs that are fertilized following spawning therefore have a single opening (micropyle) in the chorion just large enough to allow the entry of a one spermatozoon. Immediately underlying the chorion, and closely adhering to it in the unfertilized egg, is the vitelline (plasma) membrane. The central part of the egg is filled with yolk while the cytoplasm is present as a thin (~40 μ) peripheral layer that is somewhat thicker (~100 μ) at the animal pole (Ginsburg, 1968).

Meiosis in fishes, as in most chordates, is blocked at the metaphase stage of the second maturation division. The metaphase II plate and the first polar body can be found at the animal pole in close proximity to the micropyle in unfertilized eggs (Ginsburg, 1968; Brummet and Dumont, 1979) (Fig. 1). Immediately upon fertilization the egg is activated by contact between the fertilizing sperm and the ooplasm. This activation results in the completion of meiosis and the extrusion of the second polar body under the chorion. At the same time the cortical alveoli, located in the peripheral cytoplasm, discharge their colloidal contents under the chorion resulting in the formation of the perivitelline space and fluid. These changes are accompanied by migration of the peripheral cytoplasm to the animal pole, where it concentrates forming the blastodisc (Fig. 1). The formation of the fluid filled perivitelline space allows the developing egg to rotate freely within the chorion. In most eggs the blastodisc remains on top and there is no longer any link between its location and that of the micropyle. In many species, particularly salmonids, the chorion becomes a very hard protective shell (Ginsburg, 1968) (see also recent review by Hart, 1990).

Microinjection into Salmonid Eggs

Three techniques are currently being used to overcome the problem of the hard chorion of salmonids and at the same time target the nuclear areas or blastodisc. We have found that salmonid eggs [Atlantic salmon (*Salmo salar*), rainbow trout (*Oncorhynchus mykiss*), brown trout (*Salmo trutta*), and Arctic char (*Salvelinus alpinus*)] can be readily injected through the micropyle using very fine glass needles (2-3μm external diameter) (Fletcher et al. 1988, 1990, unpublished observations). In addition to facilitating microinjection procedures, the micropyle also provides us with a means of locating the male and female pronuclei. The fertilizing sperm nucleus remains in the ooplasm immediately under the micropyle in close proximity to the female pronucleus. Union of the male and female pronuclei does not appear to occur until the eggs are activated by placing them in water. Therefore microinjection through the micropyle in fertilized, but non-activated eggs places the gene constructs in very close proximity to the pronuclei.

Maclean et al. (1987) solve the problem of injecting rainbow trout eggs by using sturdy glass needles (≤15 μm) to penetrate the chorion and inject the blastodisc soon after fertilization but before the hardening process has been completed. At this time the pronuclei have fused. In the procedure used by Marc Welt of Battelle-Pacific, DNA is injected into the perivitelline space using a 33 gauge steel needle (personal communication). Rokkones et al., (1985) and Chourrout et al., (1986), who are also working with salmonids, wait until the water hardening process has been completed and then cut a small hole in the chorion to allow the insertion of a glass needle (~ 10 μm) into the well- developed blastodisc prior to first cleavage.

All of the forgoing techniques used to inject salmonid eggs result in a proportion of the fish containing what appears to be genomic integration of the injected sequences. Therefore the choice of technique will depend on the preference and strategy of the investigator. Our group prefers the micropyle method because it is the least invasive technique and we feel that the chances of an integration event prior to first cleavage are greatest if the eggs are injected as soon as possible after fertilization.

MICROINJECTION OF SALMON EGGS

Microinjection Apparatus

The microinjection apparatus (Fig. 2) consists of a glass needle driven by a micromanipulator. The needle is connected by a short length of polyethylene tubing (Clay Adams PE 200) to a "Gene Pusher" that pumps the DNA solution out of the needle using short bursts of N_2 at 200 kPa. The timing of the bursts of pressure is controlled by a power supply. The N_2 is supplied from a cylinder with the pressure value set at 30 psi.

(a) Microinjection needle

Needles are prepared from borosilicate omega dot glass tubing (1.5 mm O.D., 0.75 mm I.D., Frederick Haer and Co., Maine, USA). The glass tubing is pulled on a vertical needle puller. For injection, the needle tips are broken to 2-3 μm O.D. by touching the tip against the frosted portion of a glass microscope slide. The tip can also be broken to the required diameter using fine forceps under a dissecting microscope after the needle has been filled with DNA solution and connected to the microinjection apparatus.

The microinjection needle is mounted in a plastic holder fitted with a silicone tubing insert to prevent needle breakage, and attached to the micromanipulator. The needles are filled with DNA solution using a Hamilton (100 μl) syringe connected to Clay-Adams PE 10 polyethylene tubing. This tubing fits inside the glass needle.

The DNA solution used for injection is made up fresh daily from a concentrated stock. All buffers and distilled water used to make up the DNA solution are sterilized (autoclave) prior to use.

(b) "Gene Pusher"

This instrument is custom built by Memorial University's Department of Technical Services. An electronics technician can build one in hours at an approximate cost of $300-500 for parts.

The basic lay-out, electronic circuits and parts necessary to build the instrument are presented in figures 3-7.

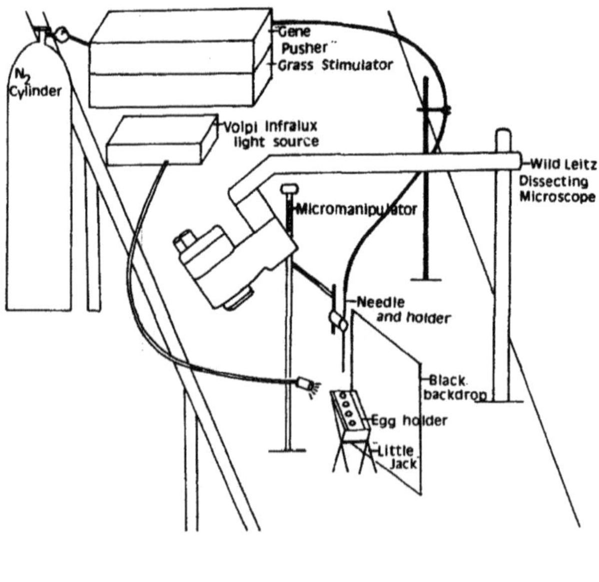

Fig. 2. Diagram of injection apparatus "set up".

FRONT VIEW OF GENE PUSHER CHASSIS

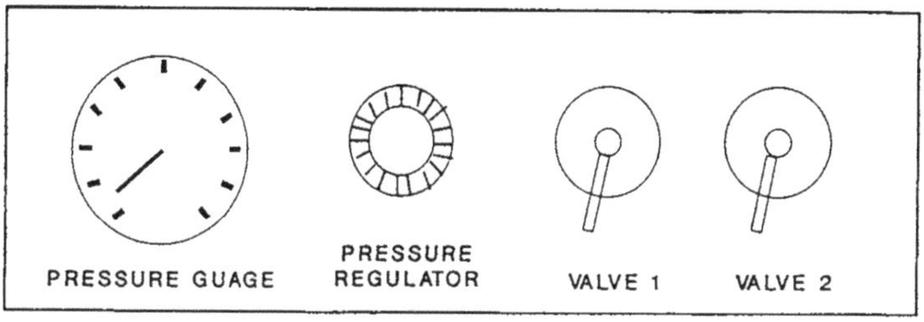

All Clippard part No.s
PARTS AND SUPPLIER
1 CLIPPARD PRESSURE GUAGE
PART NO. PG-100
1 AIR REGULATOR
PART NO. MAR1
2 TOGGLE VALVES
PART NO. MTV-4
10 STRAIGHT FITTINGS 10/32" TO 1/8"
PART NO. 15006-1
12 "O" RINGS PART NO. 11761-2
10 FT TUBING CLIPPARD PART NO. 3814-1
10 FT TUBING IMPERIAL EASTMAN
PART NO. 44P-1/4
1 CLIPPARD T FITTING PART 15002-6
6 11755 SCREW PLUGS 10/32 THREAD

SUPPLIER: WAINBEE LTD.
215 BOUL. BRUNSWICK
POINTE CLAIRE, QUEBEC
CANADA H9R 4R7

PANEL 125 ALUMINUM
PART NO. PBPA 19 007 GY2
HAMMOND MANUFACTURING

1 SOLENOID - PRESSURE REGULATOR
100 P.S.I.G.
VOLTAGE: 24 VDC
SERIAL NO.: 230
PART NO. 9-82-902

SUPPLIER: GENERAL VALVE CORP.
19 GLORIA LANE, FAIRFIELD, NEWJERSEY 07006 U.S.A.
P.O. BOX 1333 PHONE (201) 887-9460
(THIS VALVE LOCATED ON BACK PANEL, NOT SHOWN ON ABOVE DIAGRAM. SEE Fig. 4)

Fig. 3. Front view of gene pusher chassis and parts list.

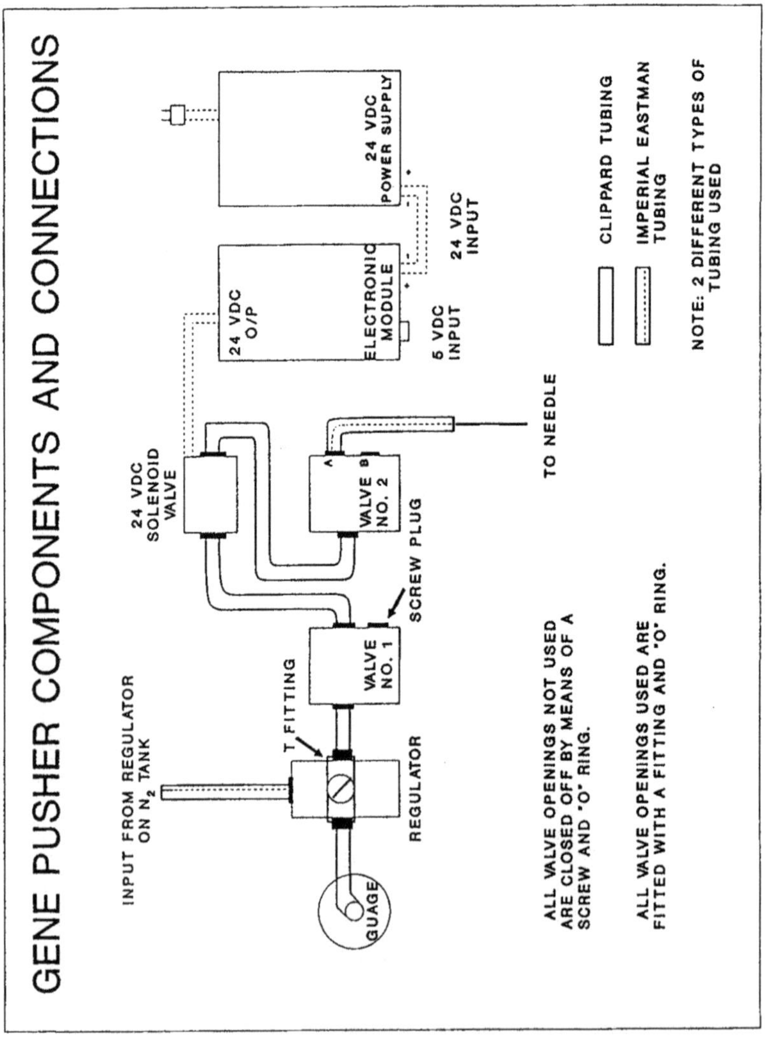

Fig. 4. Schematic of gene pusher chassis and parts list.

24 VDC POWER SUPPLY

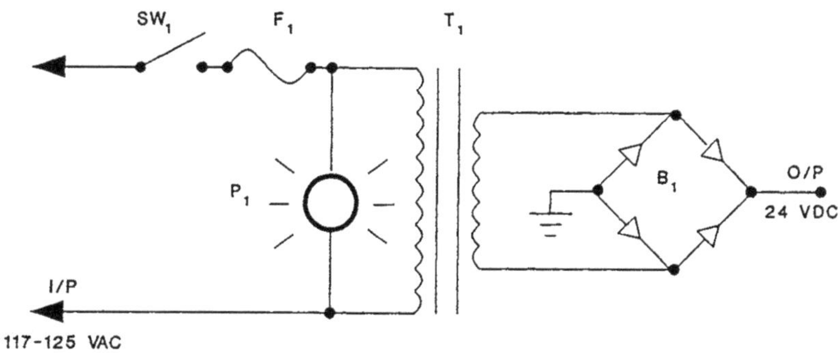

PARTS LIST

SW_1- TOGGLE SWITCH
F_1- 1.5AMP A.G.C. FUSE
P_1- 117-125 V.A.C. PILOT LAMP
T_1- HAMMOND TRANSFORMER PART NO. 167L24
B_1- BRIDGE RECTIFIER PART NO. KBPC606

ALL PARTS MOUNTED IN A HAMMOND CASE PART NO. 1411L

Fig. 5. 24 V D. C. power supply.

SCHEMATIC FOR ELECTRONIC MODULE

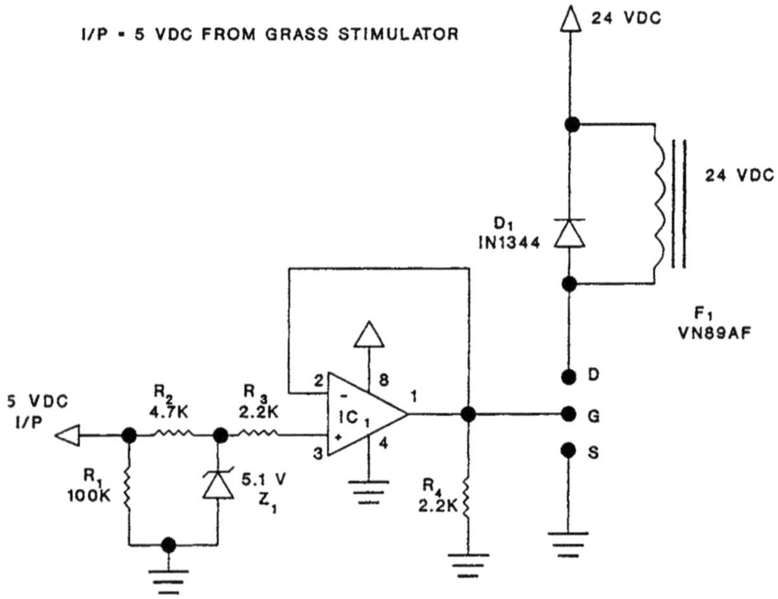

The 24 VDC power supply, supplies power to the electronic module to operate the 24 VDC solenoid valve. The FEC remains in an off state until the 5 VDC input is applied which turns on the FET allowing the solenoid to have 24 VDC on one side while completing the ground on the other. Thus 24 VDC passes through the solenoid opening the valve and allowing the N_2 to pass through it. When the 5 VDC is removed the FET is turned off and the solenoid valve closes blocking the flow of gas.

Fig. 6. Schematic for electronics module.

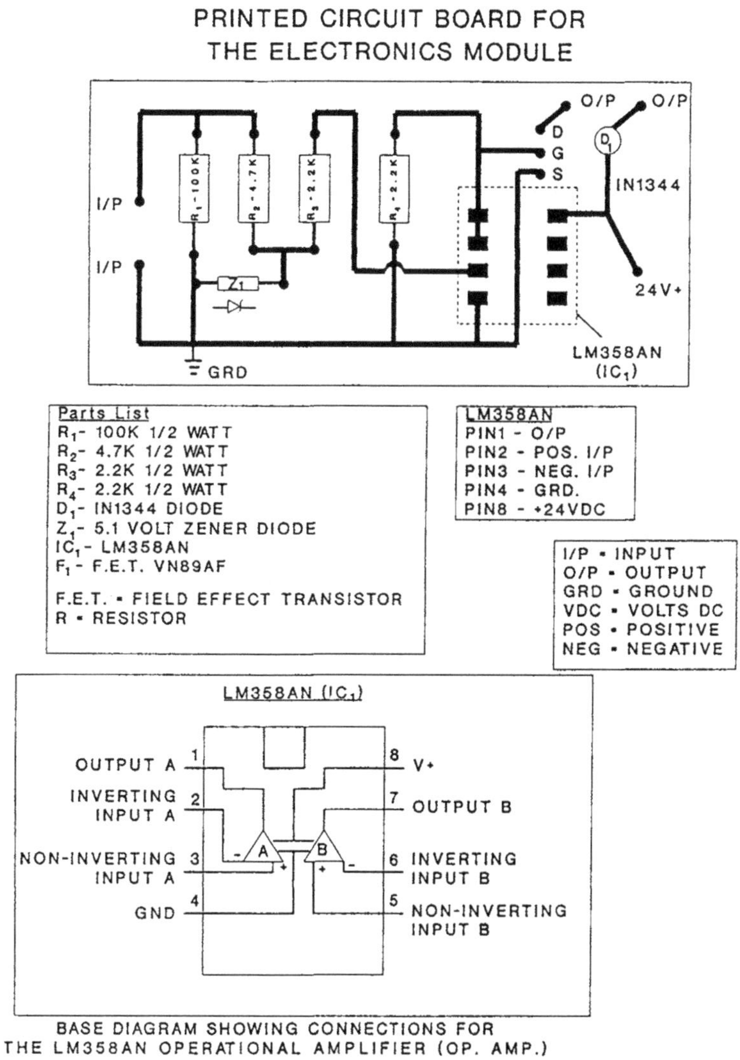

Fig. 7. Printed circuit board for the electronics module.

(c) Power supply

At the present time we are using a Grass Stimulator Model 544 to supply the power to the "Gene Pusher" and to control the frequency and duration of the bursts of N_2 pressure to the injection needle. This power supply is set to deliver one 5 VDC pulse per second with a duration ranging from 100 to 1000 milliseconds. The duration is used to control the volume (3-5 nl) of DNA pushed out of the microinjection needle. The role of the Grass Stimulator could be filled by any 5 VDC power supply with a variable pulse and duration.

Microinjection Procedures

Source of eggs

Atlantic salmon (3-5 kg) are brought in live from the field approximately 2-3 weeks before the expected spawning date (end of Oct. to mid-Nov. in Newfoundland). Male and female salmon are maintained in separate tanks supplied with flowing freshwater and air. The sexes are separated to reduce the incidence of spawning.

Collection of eggs

Eggs are stripped from ripe females which have been anaesthetized in a dilute solution of t-amyl alcohol in freshwater. Care is taken to blot the solution from the fish prior to stripping, to avoid activating the eggs. Several thousand eggs are usually collected from each fish and stored in a covered darkened container at 4°C. The eggs are usually injected within 48 hours of collection.

Sperm collection

Milt is stripped from anaesthetized male salmon, taking great care to prevent water and/or feces from contaminating the sample, since this could activate the sperm and render them useless for fertilization. One or two ml aliquots of milt are collected into 50 ml plastic beakers and covered with parafilm punctured with small holes to permit air exchange. The sperm are then stored in the dark at 4°C until needed. We have found that if stored in this manner the sperm remain viable for at least five days. It is most important that the samples are stored as thin layers and that dehydration is avoided. We usually use the samples within two days of collection.

Sperm motility test

Immediately prior to fertilization the viability of the sperm is checked using a sperm motility test. Since not all sperm samples are viable and viability can be lost during storage, it is important to carry out this test. A drop of milt is placed on a glass slide, and brought into focus under a compound microscope (200 x magnification). A drop of water is added to the milt and the sperm are judged to be viable when they exhibit rapid spiralling movements. These movements occur as soon as viable sperm come into contact with water, and they usually last no more than 30 sec.

Egg fertilization

A few drops of milt are added to several hundred "dry" eggs (eggs in ovarian fluid). The eggs and milt are gently mixed and left to sit for approximately one minute. Excess sperm is then rinsed off the eggs with several changes of cold (4°C) salmon Ringer solution (Ginsburg 1963) which had been filtered through a 0.2 μ millipore filter. Using this procedure the eggs are fertilized but not activated; activation of salmon eggs occurs when they are placed in a hypotonic solution (Ginsburg, 1968). Fertilized salmon eggs in salmon Ringer are kept on ice, in a darkened container, until injected (usually within 2 hours of fertilization).

The composition of the Ringer solution is as follows: NaCl 6.5 g, KCl 0.25 g, $NaHCO_3$ 0.2 g, $CaCl_2$ - $2H_20$ 0.4 g. All salts are dissolved in distilled water to a final volume of one litre.

Egg microinjection

The following procedures are usually carried out by two investigators acting as a team. The first investigator places four eggs in an egg holder and rotates them until the micropyle can be seen clearly on the upper surface. Once the eggs are correctly aligned they are carefully passed to the second investigator who slides the needle into the micropyle and injects the DNA. Under normal circumstances we can inject approximately 60 eggs per hour. It is important to note that the eggs are kept ice cold and bathed in Ringer solution throughout the injection procedures.

Detailed procedures

The first investigator locates the micropyles through which the second investigator injects the DNA. The micropyle of the Atlantic salmon (*Salmo salar*) consists of a micropyle pit and a micropyle canal (Riehl, 1980). The pit has a diameter of 17 μm and is 4 μm deep; the micropyle canal is 1.2 μm wide.

The first investigator transfers the fertilized eggs from the Ringer solution to an egg holder, using curved, sponge-covered forceps. The egg holder consists of a glass microscope slide covered, on the upper surface, with a layer of sponge approximately 3 mm thick. The sponge is held in place on the slide using strips of tape. The eggs are placed in small square holes cut in the sponge. The size of the holes are such that they hold the eggs firmly in place, but exert little pressure on them. The microscope slide and eggs are placed on the microscope stage perpendicular to the investigator.

The micropyle is located using a dissecting microscope (Wild Leitz) at 20 X magnification with a light source directed from the back left side of the microscope at a 45° angle to the microscope stage. The microscope and light source are enclosed in a black "felt tent" to cut down on light scatter from other sources. The position of the micropyle on the surface of the egg is located under the microscope by rotating the egg towards the light source. The light reflection around the micropyle pit differs from that of the rest of the egg surface, appearing as a darkened opaque area with a circle of reflected light around the perimeter. The egg is rotated in the holder so that the micropyle is on top, slightly back and to the left of center of the egg. When four eggs are lined up in the egg holder, a drop of Ringer is put on each egg to keep it moist.

The egg holder together with the positioned eggs are then passed to a second investigator. The egg holder is clamped with the long axis parallel to the second investigator on a small stage of adjustable height. We use a "Little Jack", purchased from Canlab, for this purpose. The "Little Jack" holding the eggs is placed in front of a black backdrop under a dissecting microscope with variable angle eyepieces (Wild M 32). A low intensity light source (fibre optics, Volpi Infralux 400) is directed at the egg from the front and at a slight angle to the egg. Background light is excluded by enclosing the microscope and eggs in a black "felt tent".

The micropyle opening is located on the surface of the egg using a combination of reflected and refracted light. To the second investigator the micropyles now appear on top and slightly forward of the centre of the egg. Each one is visualized as a black dot within an opaque area. Once the micropyle opening is located, the needle is lowered, its tip brought into focus and the "Little Jack" positioned such that the micropyle opening is directly under the needle tip. The microscope is focused on the needle tip and the micropyle. The needle is then slid into the opening. It is not possible to penetrate the egg surface (chorion) anywhere other than the micropyle with a needle tip of less than 10 μm. If the needle tip is not in the right "spot", the needle will depress the surface of the egg and break to a tip diameter of greater than 10 μm before penetration. Once the needle tip has entered the micropyle, 3-5 nl of DNA solution are injected under the cortex.

Following microinjection, the eggs are removed from the egg holder using a small spoon. They are then placed in a darkened container of fresh water and kept cold on ice (< 2h) until they are transferred to incubator trays (Heath). The eggs are incubated at 7-8°C until hatch (approximately 80 days). Dead eggs are removed from the incubator trays every few days.

COMMENTS AND DISCUSSION

We have learned over the years to be very selective about the eggs we inject. Eggs of high quality are essential for good survival. In our experience egg quality varies considerably from female to female and deteriorates with time following ovulation and following artificial stripping. We normally inject 5,000 to 10,000 eggs during a spawning season and we discard many more eggs than we inject.

We have used our standard technique, described in this manuscript, to inject a variety of linear DNA constructs (antifreeze protein, prolactin and growth hormone genes). In terms of survival to first feeding and putative integration frequencies, we have not found any obvious differences between the constructs. Our survival rates to hatch and first feeding are usually high (80%) and essentially the same as those of uninjected control eggs. However, the major reason we have high survival rates is because the investigators carrying out the injection procedures have now had approximately six years of experience. On the occasions when we have been teaching others to perform these techniques, egg survival rates have been as low as 30%. The chief difficulty encountered by the trainees is in judging how far to drive the needle into the cortex of the egg. This can only be overcome with experience.

Our long term average frequency of genomic "integration" following the injection of thousands of eggs appears to be low (2-3%), relative to those reported by Rokkones et al., (1989) (20-75%) and Guyomard et al. (1989) (38-74%). This has prompted us to design experiments to compare their injection techniques with our own.

The injection techniques used by Rokkones et al., (1989) and Guyomard et al., (1989) differ from our own in two fundamental ways. They inject after fusion of the egg and sperm nuclei and they inject considerably larger amounts of DNA (see Review by Fletcher and Davies, 1991).

Our first experiment was carried out to determine whether injecting the well-developed blastodisc would result in a higher integration frequency than we were obtaining by injecting prior to pronuclei fusion. In this experiment we fertilized the eggs in the salmon Ringer solution, orientated the micropyle on top of the egg as described in our methods, and then placed the eggs, still in the egg holder in 8° water and allowed them to be activated and water hardened. Four to five hours after water hardening (about the time when the pronuclei have fused) the eggs, still in the egg holder, were placed under the microscope and injected with 1×10^6 copies of our 7.8 kb winter flounder antifreeze gene. DNA was injected into the centre of the blastodisc through the micropyle. At this time the blastodisc is well developed and clearly visible. Six hundred salmon eggs were injected of which 500 survived to first feeding.

Analysis of DNA extracted from the red blood cells of 50 four to five month-old fry using the polymerase chain reaction (PCR) revealed that only one of them (2%) contained the injected sequence. Thus in our hands the time of injection does not influence the potential integration efficiency. This would suggest that the high integration frequencies reported by Guyomard et al., (1989) and Rokkones et al., (1989) must be attributable to some factor other than the timing of their injection; perhaps the amount of DNA injected.

There is good evidence in the literature to suggest that the frequency of extrachromosomal DNA persistence, and the length of time it persists increases with the load of DNA injected (see review by Fletcher and Davies, 1991). Thus the establishment of true genomic integration as opposed to DNA persistence becomes increasingly difficult as the injected DNA load is increased. However, there are obvious advantages to obtaining as high an "integration" frequency as possible, even if genomic integration cannot be established until the F_1 or F_2 generations. Therefore it is worthwhile investigating the relationship between DNA load and genomic integration.

We have recently initiated a series of experiments to compare the effects of injecting three different amounts of DNA (salmon GH cDNA construct) into salmon eggs: 12 pg, 200 pg and 1200 pg. The injection procedures used were identical to those described in this manuscript. These eggs are just beginning to hatch, and DNA integration frequencies have not yet been determined. However it is clear that large amounts of DNA are not toxic to the embryos because survival to hatch did not differ between the three groups.

In a recent study Inoue et al., (1991) reported that glutathione (reduced) prevented the water hardening of freshly fertilized rainbow trout eggs, thus allowing the chorion to be readily penetrated by a fine glass injection needle. We have tried to use this technique in Atlantic salmon eggs. However, although hardening of the eggs was inhibited, microinjection remained very difficult. Indeed, we found that microinjection of ovulated, unfertilized eggs through the chorion was difficult. At the present time we cannot explain why Atlantic salmon and rainbow trout eggs appear to differ. It may be that the chorions of "wild" stock Atlantic salmon eggs are considerably more robust than those of artificially cultured rainbow trout.

ACKNOWLEDGMENTS

We are grateful to Vern Pepper, Chuck Bourgeois and John Gibson, Department of Fisheries and Oceans, Canada, for providing us with assistance and Atlantic salmon. Thanks are also due to Valda Boland for keeping our young salmon healthy and to Terry Harris, Memorial University Technical Services for constructing the "gene pusher". This research is supported by the Natural Sciences and Engineering Research Council of Canada. Ocean Sciences Centre Contribution Number 129.

REFERENCES

Brinster, R.L., Chen, H. Y., Trumbauer, M. E., Yagle, M. K. and Palmiter, R. D. (1985) Factors affecting the efficiency of introducing foreign DNA into mice by microinjecting eggs. Proc. Natl. Acad. Sci. USA 82: 4438-4442.

Brummet, A.R. and Dumont, J. N. (1979) Initial stages of sperm penetration into the egg of *Fundulus heteroclitus*. J. Exp. Zool. 210: 417-433.

Chourrout, D., Guyomard, R. and Houdebine, L. M. (1986) High efficiency gene transfer in rainbow trout (*Salmo gairdneri* Rich.) by microinjection into egg cytoplasm. Aquaculture 51: 143-150.

Du, S. J., Gong, Z., Fletcher, G. L. Shears, M. A. and Hew, C. L. (1992) Growth hormone gene transfer in Atlantic salmon: use of fish antifreeze/growth hormone chimeric gene constructs. In Transgenic Fish. C. L. Hew and G. L. Fletcher (eds.) World Scientific Publications, Singapore, pp. 176-189.

Fletcher, G.L. and Davies, P. L. (1991) Transgenic fish for aquaculture. In: Genetic Engineering, Principles and Methods. Jane Setlow (ed.) Plenum Press. 13: 331-370.

Fletcher, G.L., Shears, M. A., King, M. J., Davies, P. L. and Hew, C. L. (1988) Evidence for antifreeze protein gene transfer in Atlantic salmon (*Salmo salar*). Can. J. Fish. Aquat. Sci. 45: 352-357.

Fletcher, G.L., Shears, M. A., King, M. J., Kao, M. H., Davies, P. L. and Hew, C. L. (1990) In Fish Physiology, Fish Toxicology and Fisheries Management: Proceedings of an International Symposium, Guangzhou, PRC, September 14-16, 1988 (Ryans, R.C. ed). United States EPA. Environmental Research Laboratory, Athens, GA. EPA/600/9-90/011. pp. 230-247.

Ginsburg, A.S. (1963) Sperm-egg association and its relationship to the activation of the egg in salmonid fishes. J. Embryol. Exp. Morph. 11: 13-33.

Ginsburg, A.S. (1968) Fertilization in fishes and the problem of polyspermy. Academy of Sciences of the USSR, Institute of Developmental Biology. Translated from Russian by Z. Blake and edited by B. Golek. (Israel program for Scientific Translations, Jerusalem 1972, Keter Press Binding: Wiener Binding Ltd. Jerusalem. 366p.).

Guyomard, R., Chourrout, D., Leroux, C., Houdebine, L. M. and Pourrain, F. (1989) Integration and germ line transmission of foreign genes microinjected into fertilized trout eggs. Biochemie 71: 857-863.

Hammer, R.E., Pursel, V. G., Rexroad Jr., C. E., Wall, R. J., Bolt, D. J., Ebert, K. M., Palmiter, R. D. and Brinster, R. L. (1985) Production of transgenic rabbits, sheep and pigs by microinjection. Nature (London) 315: 680-683.

Hart, N. H. (1990) Fertilization in teleost fishes: mechanisms of sperm-egg interactions. Int. Rev. Cytol. 121: 1-66.

Inoue, K., Yamashita, S., Akita, N., Mitsuboshi, T., Nagahisa, E., Shiba, T. and Fujita, T. (1991) Histochemical detection of foreign gene expression in rainbow trout. Nippon Suisan Gakkaishi. 57(8): 1511-1517.

Maclean, N., Penman, D. and Talwar, S. (1987) Introduction of novel genes into the rainbow trout. Proc. World Symp. on Selection, Hybridization and Genetic Engineering in Aquaculture. Bordeaux, May 27-30, 1986, Vol. II, pp 325-333 Berlin.

Riehl, R. (1980) Micropyle of some salmonins and coregonins. Environ. Biol.Fishes. 5: 59-66.

Rokkones, E., Alestrom, P. Skjervold, H. and Gautvik, K. M. (1985) Development of a technique for microinjection of DNA into salmonid eggs. Acta Physiol. Scand. 124 (Suppl 542): 417.

Rokkones, E., Alestrom, P., Skjervold, H. and Gautvik, K. M. (1989) Microinjection and expression of a mouse metallothionein human growth hormone fusion gene in fertilized salmonid eggs. J. Comp. Physiol. 158B: 751-758.

Strojek, R.M. and Wagner, T.E. (1988) The use of transgenic animal techniques for livestock development. In Genetic Engineering, Principles and Methods. Jane Setlow (ed.) Plenum Press. 10: 221-246.

Yamaha, E., Onozato, H. and Yamazaki, F. (1988) Visualization of female pronucleus in the goldfish (*Carassius auratus*) using fluorescent dye, hoechst 33342. Nippon Suisan Gakkaishi 54: 537.

CHAPTER 4

The Application of PCR to Aquaculture

Peter L. Davies and Sherry Y. Gauthier

ABSTRACT

The advent of the polymerase chain reaction (PCR) has greatly simplified screening for specific DNA sequences, a task which was previously done by Southern blotting. There are numerous applications of PCR to aquaculture where the ability to screen large numbers of samples rapidly and at low cost is particularly important. This article contains methods for the preparation of DNA from fish tissues, blood and milt, and its application in PCR reactions. Some precautions are outlined that should help prevent DNA contamination giving rise to false positives. The use of an internal control to detect unamplifiable DNA samples is also described. The specific example used to illustrate the application of PCR in aquaculture is the detection of flounder antifreeze protein gene transfer to Atlantic salmon.

INTRODUCTION

The polymerase chain reaction (PCR) was designed to amplify a specific DNA sequence between two oligonucleotide primers (Saiki et al., 1985). By this method it is possible to detect as little as one copy of the target DNA out of a complex DNA background. The power and simplicity of the technique has led to its rapid acceptance by the life science community. Subsequently, PCR has been put to many creative uses, as for example in selective cDNA cloning by the rapid amplification of cDNA ends (RACE) protocol (Frohman et al., 1988), and the inverse PCR amplification to detect and isolate sequences flanking a DNA of interest (Ochman et al., 1988) etc. It is not the purpose of this article to cover the principles of the method or to provide an overview of the area, a function which is well served by a number of excellent reviews (Erlich et al., 1988; Marx, 1988; Oste, 1988). Rather, it is intended to give some practical examples, methodology, and advice on the techniques as applied to aquaculture. Some of the applications in this area are; 1) stock identification, particularly where genetically manipulated or tagged strains are being used; 2) screening for the passage of genes in a breeding program; 3) sexing fish; 4) screening for viral and bacterial infections.

Our experience with PCR has revolved around gene transfer into fish and the segregation of the gene in subsequent generations (Fletcher et al., 1988b; Shears et al., 1991). Initially, the genes of interest were those coding for antifreeze protein (AFP) from winter flounder (Davies et al., 1989). The objective of the gene transfer is to confer to Atlantic salmon, which have no natural antifreeze (Fletcher et al., 1988a), the same level of freeze-resistance shown by flounder. If freeze-resistant salmon were available they could be farmed successfully on the East coast of North America where winter conditions of ice in seawater would otherwise cause massive mortality. A more extensive discussion on the biology of macromolecular antifreezes and antifreeze protein gene transfer is presented by Fletcher et al., (1992) (Chapter 11 of this monograph).

MATERIALS AND EQUIPMENT

An attractive feature of PCR methodology is that it does not require expensive equipment. A thermocycler, horizontal gel electrophoresis apparatus, and UV light are virtually all that is needed to equip a lab. This makes it feasible to have an on-site PCR capability and avoid sample shipment to a central location. However, strenuous precautions must be taken to avoid cross-contamination of stock solutions and samples, which requires cleanliness and good working practices, and perhaps even a dedicated room/area to be used only for PCR.

A variety of automatic thermocyclers are available ranging in price from $3000 to $8000. The less expensive machines are either cooled by water or a Pelletier junction. The latter has a finite life-span and must be replaced eventually, at which point the sample block needs to be recalibrated. It is a good idea to have a spare junction on hand. The water-cooled machines have a slight inconvenience in that they cannot hold a temperature below that of the cooling water. There are also the uncertainties of plumbing to consider. The instrument we use for PCR is a water-cooled Techne® PHC-2 programmable temperature cycler with sample block for 54 x 0.5 ml microcentrifuge tubes. Gel photographs are taken with a Polaroid MP4 land camera using illumination from a Fotodyne UV light.

Operating costs are also reasonable. Of the expendable items, the *Thermophilus aquaticus* (Taq) heat-stable DNA polymerase (Promega) and oligonucleotides are the most expensive. The relative cost of the oligonucleotides decreases with large-scale synthesis or purchase as would occur with an established assay. The opportunity exists to scale down the volume of the reaction since so little product is required, even for visual inspection by ethidium bromide staining. Also, there is the potential to automate certain aspects of the assay to handle large numbers of samples.

METHODS

Method I: Extraction of high molecular weight DNA from whole tissue

This method is based on the procedure of Blin and Stafford (1976) and is ideal for the preparation of high molecular weight DNA suitable for genomic Southern blotting and genomic library construction.

Tissue (4g) is quick frozen in liquid N_2 and ground to a fine powder under liquid N_2 in a precooled pestle and mortar. In the same mortar 40 ml of extraction buffer (0.1M Tris-base, 0.05 M Na_2 EDTA, 0.2 M NaCl, 1% SDS containing 100 µg/ml proteinase K) is frozen in a plastic weighing boat floated on liquid N_2 and then ground to a fine powder while being mixed in with the pulverised tissue. The mixture is transferred to a conical flask in a rotary incubator and swirled gently at 60-65°C with a further addition of 100 µg/ml proteinase K after 1-3h. After 12-16h the extract should be homogeneous and show evidence of viscosity. If tissue clumps remain it is an indication that the grinding and mixing were incomplete. Proteinaceous impurities are removed from the DNA solution by three successive extractions with equal volumes of phenol/chloroform (50:50). The phenol should be prepared and equilibrated as described by Maniatis et al., (1982). The aqueous phase containing the DNA is dialysed against 100 volumes of 10 mM Tris-HCl (pH 8.0), 1 mM EDTA (TE buffer), with three changes over 24h. The DNA solution can be concentrated by secondary-butanol extraction followed by dialysis against TE buffer.

This protocol is laborious but produces good quality DNA. Cruder DNA preparations are generally suitable for PCR analyses. If, however, a DNA sample fails to amplify, this method can be used to upgrade the DNA purity. Typical, but generally innocuous, contaminants are glycogen and RNA. The former can largely be removed by ultracentrifugation and the latter by digestion with DNAse-free RNAse followed by phenol/chloroform extraction and extensive dialysis (as above). Virtually any soft tissue is suitable for DNA extraction but maturing testes are a particularly rich source of DNA, so much so that it is necessary to reduce the ratio of tissue to extraction solution by as much as an order of magnitude. Otherwise the mixture becomes too viscous for an effective extraction.

(100 μl) without the need for tissue grinding. After dilution in 100 mM Tris-HCl (pH 8.0), 40 mM EDTA (2 ml) the cells are lysed by mixing with 2 ml of extraction buffer (see Method I). The mixture is incubated at 60-65°C overnight and the extraction continued as in Method I.

Method III: Release of crude DNA from blood or sperm

A simple method for accessing DNA for PCR is to place ~1 μl of whole blood, a blood cell pellet after plasma removal, or milt in 50 μl of 10 mM NaOH. The mixture is boiled for 5 min and then spun for 1 min in a microfuge to pellet cell debris. This extraction can be scaled down because only 1-2 μl of the supernatant fraction is needed for a typical PCR reaction. The method can be adapted to release "PCR-quality" DNA from tissue pieces such as whole embryos by extending the incubation time at 100°C to 10 min and by increasing the amount of NaOH solution used.

Method IV. PCR reaction

Typically the PCR reaction is done in a 500 μl microcentrifuge tube in a final volume of 50 μl containing 10 mM Tris-HCl (pH 9.0), 50 mM KCl, 1.5 mM $MgCl_2$, 0.1% Triton X-100, ≤ 1 μg of template DNA, 20 pmol of each oligonucleotide primer, 10 nmol of each deoxynucleoside triphosphate and 2.5-5 units of Taq DNA polymerase (Promega). The reaction mixture is covered with 35 μl of mineral oil before loading into the PCR machine. Each cycle of amplification involves:

1) denaturation of DNA at 95°C for 1 min
2) primer annealing at 60°C for 1 min
3) primer extension at 72°C for 2 min

After 25 cycles, 10 μl of the reaction product is removed from under the mineral oil layer, mixed with 2 μl of 5x dye solution (30% glycerol, 15% Ficoll, 0.02% bromophenol blue, 0.02% xylene cyanol FF in TE buffer) and analysed by electrophoresis on a 2% agarose gel alongside a 100 bp or 123 bp DNA ladder. The gel is stained in ethidium bromide (10 μg/ml) and viewed or photographed under UV light.

The variables that can be altered in the PCR reaction are the denaturation temperature, which should be higher for a G.C-rich target DNA than an A.T-rich DNA, and the annealing temperature, which depends on the melting temperature and length of the oligonucleotides used. If the annealing temperature is set too low for the primer set used then many non-specific products can result. The optimum annealing temperature can be determined by using 50°C for the initial experiment and then gradually increasing the temperature until only the one band of interest remains.

used then many non-specific products can result. The optimum annealing temperature can be determined by using 50°C for the initial experiment and then gradually increasing the temperature until only the one band of interest remains.

EXAMPLES AND DISCUSSION

In setting up a PCR analysis, as with any experiment, it is important to have appropriate controls. The two main pitfalls to guard against are contamination of stocks, particularly by the PCR product being screened for, which generates false positives, and the failure of the amplification reaction, which could be scored as a false negative.

Contamination

This is widely believed to be caused by DNA-containing aerosols that are produced on opening microcentrifuge tube lids, ejecting tips from micropipettors, and suctioning of samples into the tips. Even the hands and clothing of the worker are suspect. To guard against contamination it is essential to include a control reaction containing all the ingredients except the template DNA (Fig. 1 A, lane 11). If this control shows a positive reaction it is possible to track down the source of the contamination by performing PCR reactions with various permutations of old and new, contamination-free reagents. In practice it is usually simpler to just start over with new solutions. For this reason it is wise to aliquot the expensive reagents like Taq polymerase and oligonucleotide primers before they are brought into the PCR facility, using pipettes and equipment that have not been associated with the DNA of interest. An alternative is to irradiate the reagents with UV light to cross-link and inactivate any DNA contaminants present (Sarkar and Sommer, 1990). Naturally, this cannot be done for the template DNA. Indeed there is no simple way of detecting or correcting contamination of the target DNA. The Taq polymerase is also inactivated by UV light. An excellent discussion of the dangers of PCR contamination and how to avoid them is given by Kwok and Higuchi (1989).

Among the precautions that we routinely observe are the use of an isolated area for dispensing PCR reagents prior to the addition of DNA, as well as a separate set of pipettors, tips, tubes, racks and gloves etc. When all other PCR reagents are present in the reaction tubes, including the mineral oil, the closed tubes are brought into the main lab. DNA is dispensed using a positive displacement pipettor or a disposable micropipette. Each tube is capped immediately after the addition of DNA and before proceeding to the next sample. The positive control is pipetted last and whenever possible this is a control that is similar to the samples in its concentration of target DNA sequence. If an enriched source has to be used, such as a plasmid, it should be highly diluted.

No Product

To control for the failure of a PCR reaction appearing as a false negative it is useful to coamplify a second sequence which is known to be present in the template DNA. When screening salmon for the incorporation and inheritance of the flounder AFP gene by PCR we routinely include a primer set that will amplify a section of the salmon growth hormone gene (Fig. 1B). These primers (Table I) amplify sequences

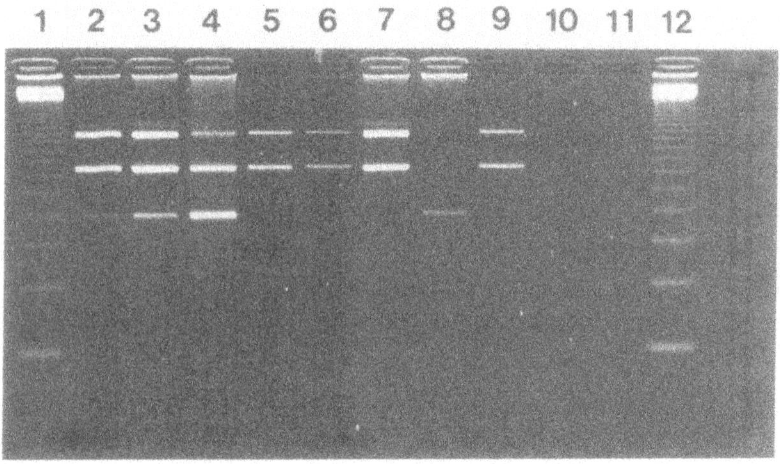

Fig. 1A. Detection of AFP gene transfer by PCR. Reaction products (10 μl) from 25-cycles of amplification were analyzed by electrophoresis in a 2.0% agarose gel stained with ethidium bromide. The reactions analyzed in lanes 2, 3, 4, 5, 6 and 11 were done with both the growth hormone gene and the AFP gene primer sets present. The reactions analyzed in lanes 7 and 9 received only the growth hormone gene set and those analyzed in lanes 8 and 10 received only the AFP gene set. Salmon template DNAs were as follows: lane 2), genomic DNA from sperm of fish #1469; lane 3) genomic DNA from fish #1231; lane 4) lysed blood from fish #1457; lane 5) genomic DNA from fish #24; lane 6) lysed blood from fish #ASC4; lanes 7 and 8) genomic DNA from sperm of fish #1469; lanes 9 and 10) genomic DNA from fish #24; lane 11, no DNA (control); lanes 1 and 12, 123 bp DNA ladder. (By permission of Blackwell Scientific Publications, Inc.)

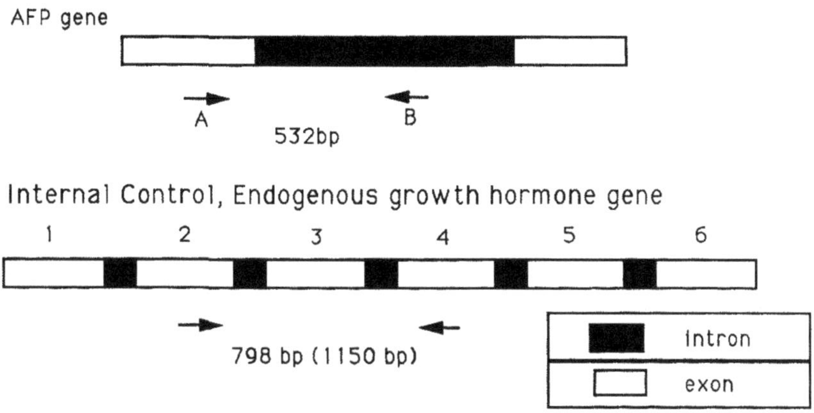

Fig. 1B. Schematic diagram showing the positioning of primer sets on the antifreeze protein and growth hormone genes.

Table I **PCR primers**

	Antifreeze protein gene	
5'	ATGGCTCTCTCACTTTTCACTGT	3'
position 50	(sense)	
5'	CACTTTCACCAAACTCAGTGATG	3'
position 532	(antisense)	
	Salmon growth hormone gene	
5'	GCTCTTCAACATCGCGGTCA	3'
position 607	(sense)	
5'	ATATGGAGCAGCTTCAGGAC	3'
position 1404	(antisense)	

spanning from bases 607 to 1404 in the second and fourth exons, respectively, of the growth hormone gene (Johansen et al., (1989)). In addition to this 798 bp product a second DNA band is consistently seen at ~1150 bp (Fig. 1 A, lanes 2-7 and 9) in all amplifications using the growth hormone primer set. This product is presumed to be derived from a second copy of the growth hormone gene, equivalent to that seen in the rainbow trout (Agellon et al., 1988). The difference in length of these products could be due to a difference in the size of the second and/or third introns.

When small volumes of blood are freshly prepared for PCR by method III they almost invariably amplify successfully. However, we have been less successful with blood samples that have been frozen and thawed a number of times. In this situation it is best to obtain another fresh sample.

Screening for Antifreeze Protein Gene Transfer

The primers used to detect the AFP gene in transgenic salmon (Table I) amplify a section from base 50 of the 5'-exon to base 532 of the intron (Scott et al., 1988) to give a 483 bp PCR product (Fig. 1 A, lane 8, Fig. 1 B). When both primer sets are used to amplify DNA from salmon that are known to be transgenic for the winter flounder AFP gene by Southern blotting, all three PCR products are present (Fig. 1 A lanes 2 and 3). Conversely, when both primer sets are used with DNA from salmon that were shown by Southern blotting not to be transgenic, only the two growth hormone-specific bands appear (Fig. 1 A, lanes 5 and 6). With this assay it is easy to identify an unknown (fish #1457) as transgenic (Fig. 1 A, lane 4), but it is equally easy to pick out non-transgenic fish because of the internal control provided by the growth hormone primers. Thus in Fig. 2, where blood from F1 individuals have been screened for the transmission of the AFP gene, ASC 300, 302, 304, 306, 308, 309, 313 and 314 on the top row are clearly non-transgenic.

It is feasible for one person working with one PCR machine to process and analyse approximately 100 samples/day. This is an improvement in efficiency of two orders of magnitude over the conventional Southern blotting procedure, with an attendant increase in sensitivity and a considerable decrease in the cost per sample.

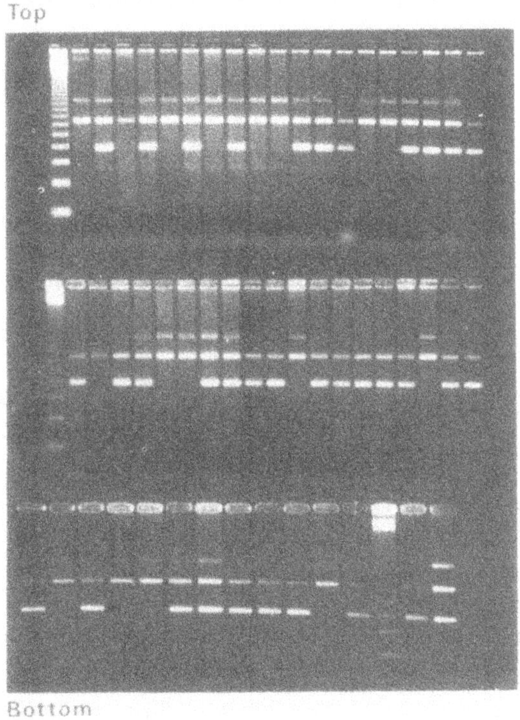

Fig. 2. Transmission of the AFP transgene to F_1 individuals. Blood from 50 progeny (ASC 300-349) of the cross between male #1542 and a wild-type female were analysed for the presence of the AFP gene by PCR. Reaction products were examined as described in the legend to Fig. 1. Top row (L to R): standard 123bp DNA ladder, ASC 300 to 318. Middle row (L to R): standard 123bp DNA ladder, ASC 319 to 337. Bottom row (L to R): ASC 338 to 349, standard 123bp DNA ladder, blood from ASC 178 (positive control), genomic DNA from the sperm of F_0 fish #1469, no template DNA (negative control). (By permission of Blackwell Scientific Publications, Inc.)

ACKNOWLEDGMENTS

Equipment and research funds were obtained from the Medical Research Council of Canada and from the National Sciences and Engineering Research Council of Canada strategic grants program. PCR primers were made by the Oligonucleotide Synthesis Laboratory, Queen'sUniversity. We are grateful to the Editor of Marine Molecular Biology and Biotechnology for permission to reproduce figures 1 and 2 from Shears et al., (1991).

REFERENCES

Agellon, L.B., Davies, S.L., Chen, T.T. and Powers, D.A. (1988) Structure of a fish (rainbow trout) growth hormone gene and its evolutionary implications. Proc. Natl. Acad. Sci. U.S.A. 85: 5136-5140.

Blin, N. and Stafford, D.W. (1976) A general method for isolation of high molecular weight DNA from eukaryotes. Nucleic Acids Res 3: 2303-2308.

Davies, P.L., Fletcher, G L. and Hew, C.L. (1989) Fish antifreeze protein genes and their use in transgenic studies. Oxford Surveys on Eukaryotic Genes . Maclean, N. (ed.) Oxford University Press, 6: 85-109.

Erlich, H.A., Gelfand, D.H. and Saiki, R.K. (1988) Specific DNA amplification. Nature 333: 461-462.

Fletcher, G.L., Kao, M.H and Dempson, J.B. (1988a) Lethal freezing temperatures of Arctic char and other salmonids in the presence of ice. Aquaculture 71: 369-378.

Fletcher, G.L., Shears, M.A., King, M.J., Davies, P.L. and Hew, C.L. (1988b) Evidence for antifreeze protein gene transfer in Atlantic salmon (*Salmo salar*). Can. J. Fish. Aquat. Sci. 45: 352-357.

Fletcher, G. L., Davies, P. L. and Hew, C. L. (1992) Genetic engineering of freeze-resistant Atlantic salmon. In: Transgenic fish. C. L. Hew and G. L. Fletcher (eds.) World Scientific Publishing Co., Singapore. pp. 190-208.

Frohman, M. A., Dush, M. K. and Martin, G. R. (1988) Rapid production of full-length cDNAs from rare transcripts: Amplification using a gene-specific oligonucleotide primer. Proc. Natl. Acad. Sci. U.S.A. 85: 8998-9002.

Johansen, B., Johnson, O. C. and Valla, S. (1989) The complete nucleotide sequence of the growth hormone gene from Atlantic salmon (*Salmo salar*). Gene 77: 317-324.

Kwok, S. and Higuchi, R. (1989) Avoiding false positives with PCR. Nature 339: 237-238.

Maniatis, T., Fritsch, E.F. and Sambrook, J. (1982) Molecular cloning: a laboratory manual. Cold Spring Harbor Laboratory, NY.

Marx, J.L. (1988) Multiplying genes by leaps and bounds. Science 240: 1408-1410.

Ochman, H., Gerber, A.S. and Hartl, D.L. (1988) Genetic Applications of an inverse polymerase chain reaction. Genetics 120: 621-625.

Oste, C. (1988) Polymerase chain reaction. Biotechniques 6: 162-167.

Saiki, R.K., Schaif, S., Faloona, F., Mullis, K.B., Horn, G.T., Erlich, H.A. and Arnheim, N. (1985) Enzymatic amplification of ß-globin genomic sequences and restriction site analysis for diagnosis of sickle cell anemia. Science 230: 1350-1350.

Sarkar, G. and Sommer, S.S. (1990) Shedding light on PCR contamination. Nature 343: 27.

Scott, G.K., Davies, P.L., Kao, M.H. and Fletcher, G.L. (1988) Differential amplification of antifreeze protein genes in the *Pleuronectinae*. J. Mol. Evol. 27: 29-35.

Shears, M.A., Fletcher, G.L., Hew, C.L., Gauthier, S. and Davies, P.L. (1991) Transfer expression, and stable inheritance of antifreeze protein genes in Atlantic salmon. Mol. Marine Biol. and Biotech. 1: 58-63.

CHAPTER 5

Medaka and Zebrafish: Ideal as Transient and Stable Transgenic Systems

Juergen Robert Vielkind

ABSTRACT

Methodology for transient expression and stable inheritance of transgenes in medaka and zebrafish is described. DNA is cytoplasmically injected into fertilized eggs preferably prior to cleavage of the egg. Injected DNA in linear and supercoiled form or transferred in phage particles is replicated, degraded but also retained in a fraction of young fish. The DNA is transiently expressed but, after mosaic integration into the germ line of 5% of injected fish, is stably intregrated into the chromosomes, expressed in the progeny and inherited in a Mendelian fashion. Although with one exception only heterologous transcriptional control regions have been analysed, it appears that the reporter genes are transiently expressed as well as in the progeny in a tissue specific manner.

In summary, both the medaka and zebrafish are ideal systems to study the function of genes governing cell development and differentiation.

INTRODUCTION

The introduction of methodology for transferring pure genes (transgenes) by microinjection into oocytes (Brown and Gurdon, 1977), fertilized eggs (Gordon et al., 1980) or the developing embryos (Schnieke et al., 1988) has widely expanded our ability to study regulated spatial and temporal gene expression and to isolate desirable genes (Allen et al., 1988) or their control regions (Jaenisch and Mintz, 1974) via insertion of tag sequences. The technique, however, also bears the potential to manipulate the vertebrate genome and has resulted in the introduction of new traits into animals of medical (Sinn et al., 1987) and economic (Hammer et al.,1985) interest.

The transfer of genes and the production of transgenics have been particularly well established in the mouse (reviewed in Palmiter and Brinster, 1986) and fruitfly (Spradling and Rubin, 1982; Rubin and Spradling, 1982). Application of these techniques to fish dates back only a few years and has been mainly concerned with problems in fisheries aimed at improving broodstocks of economically important fish (Chourrout et al., 1986). Therefore, in contrast to the studies in other animals which centered on basic biological problems, in fish the basics such as integration, expression, and germline inheritance of transgenes are not yet resolved. In addition, only a very few fish genes and transcriptional control regions have been isolated and therefore most of the studies to date have been performed using genes of higher vertebrates (reviewed in Ozato et al., 1989).

GENERAL FEATURES

Fish are ideal for gene transfer studies. Most fish species are egg-layers with external fertilization so that many eggs can easily be obtained on a daily basis and the embryos reared to adulthood. Many manipulations that are necessary in higher vertebrates, e.g. culture of eggs, transfer to foster mothers after injection of the transgene, are therefore not necessary. However, large yolk masses make the localization of the usually small zygote nucleus difficult which in many cases is made even more difficult by an opaque chorion. Therefore injections are done into the perinuclear cytoplasm (Chourrout et al., 1986). Alternatively, the DNA is injected into the clearly visible germinal vesicle (Ozato et al., 1986, 1992, Chapter 2, this monograph) of the unfertilized egg. This circumvents another problem, the tough chorion which hinders microinjection. Various techniques have been developed to overcome this obstacle such as digestion with proteinases, manual removal of the chorion, or boring holes into it (reviewed by Maclean et al., 1987; Fletcher and Davies, 1991).

Requirements for and Choice of Models

In higher vertebrates the whole of gene transfer technology has been based on the mouse as a model (reviewed in Palmiter and Brinster, 1986). Until recently (Ozato et al., 1986; Stuart et al., 1987; Chong and Vielkind, 1989; Inoue et al., 1989; Vielkind et al., 1990; Stuart et al., 1990; Winkler et al., 1991) such models have not been in existence in fish. To establish a fish model, DNA application techniques, DNA fate and integration as well as expression and stable germ-line transmission of transgenes must be investigated. In addition, fish models should feature a transparent chorion, easy injection and, important for inheritable transgene studies, a short generation time combined with year round breeding.

Two fresh water fish, the Japanese medaka (*Oryzias latipes*) and the zebrafish (*Brachydanio rerio*), represent ideal candidates for such models. Their biology, including embryonic development and physiology, have been extensively studied (Yamamoto, 1967; Laale, 1977); they have a short embryogenesis, medaka hatch in 10-14 days and zebrafish in 2-6 days after fertilization and are sexually mature after 2-4 months; comparable to that of mice. Both are egglayers producing hundred of eggs on a regular basis, do not exhibit seasonal breeding behaviour, and can be maintained very easily. Their eggs are relatively large (diameter: 1.5mm for medaka and 0.7mm for zebrafish) and possess relatively thin and transparent chorions. These features allow injection without mild pronase digestion if the proper needle is used and make the first cell of the fertilized egg clearly visible for positioning of the injection needle into its cytoplasm. Finally, inbred stocks of both species are available (Streisinger et al., 1981; Hyodo-Taguchi and Egami, 1985). This is of importance because it allows evaluation of the expression of transgenes in a defined genetic background.

PROCEDURES

Fish Breeding, Egg Collection and Embryo Culture

Medaka: To induce spawning and fertilization the usual 14hr light/10hr dark cycle is changed to a 16hr light and 8hr dark period. During the light period the temperature is kept close to 27°C but a slight temperature decrease during the dark phase and 3 sparse feedings daily increase egg production. A ratio of 30 females to 10-15 males in a 40 liter tank also facilitates breeding. Under these conditions a female produces 10-30 eggs almost every day during a 4-6 week period 1-2hr after onset of light. After this time the fish should be given a rest of 1-2 months by being kept at the normal light-dark cycle and if possible at lower temperatures.

The fertilized eggs are attached to the female in clusters and can easily be removed with fine forceps. Until injection the eggs are kept in Ringer solution (0.75% NaCl, 0.02% KCl, 0.02% $CaCl_2$, pH 7.3) at 12°C to slow down cleavage processes and hardening of the chorion. After injection they are reared in 0.1% NaCl, 0.003% KCl, 0.004% $CaCl_2 \cdot 2H_20$, 0.0163% $MgSO_4 \cdot 7H_20$, 0.0001 % methylene blue and finally transferred to aquarium water after hatching (for details, see Matsui, 1949; Yamamoto, 1967; Chong and Vielkind, 1989; Vielkind et al., 1990).

Zebrafish: In contrast to medaka, zebrafish should only be used for embryo collection twice a week with a one week resting interval due to the larger number of eggs produced. Therefore, females and males are kept in separate tanks at 27°C in a 14hr light/10hr dark phase. Eight females together with 4 males are placed in a 10-20 liter tank the evening before embryos are desired. To prevent the fish from eating the freshly spawned eggs the bottom of the tank is covered by one layer of marbles. The fertilized eggs can then be easily siphoned off the bottom during the 1-2 hr spawning period following the onset of light. Collected eggs are transferred to a modified Hanks (10% containing 1.3mM $CaCl_2$, 1 mM $MgCl_2$, 4mM $NaHCO_3$) in which they are kept until hatching. The fish are then transferred to normal tank water (for details, see Laale, 1977; Stuart et al., 1987, 1990).

DNA Application: Microinjection

There are a number of ways to transmit DNA into cultured cells, $CaPO_4$ DNA co-precipitation, electroporation, microinjection and the use of carriers such as liposomes or modified viral particles. For the transmission of exogenous DNA into fish oocytes or fertilized eggs microinjection techniques have been employed exclusively until recently when electroporation has been reported as another successful method (Inoue et al., 1990). A shortcoming of this method, however, is that only 25% of treated eggs survive to the hatching stage.

Whereas in the mouse (Brinster et al., 1985) pronuclear injection is necessary for survival, integration and function of the foreign DNA, in other organisms such as fruitflies (Spradling and Rubin, 1982; Rubin and Spradling, 1982), nematodes (Kimble et al., 1982), sea urchins (McMahon et al., 1985) and frogs (Rusconi and Schaffner, 1981) cytoplasmic injections have been successfully employed. Therefore, in the present study cytoplasmic microinjection has been used with fertilized medaka and zebrafish eggs. For microinjection fertilized medaka eggs are placed in egg-size wells made in a 1.5% agarose layer filled with Ringer solution, zebrafish eggs are placed in a ringer filled depression slide. Due to the hard chorion it is best that zebrafish eggs are dechorionated manually or by digestion with pronase (500µg/ml) (Stuart, 1990). To avoid damage to the embryo a thorough washing in modified Hanks solution is necessary. Cytoplasmic microinjection is done before first cleavage under a stereomicroscope and with a simple micromanipulator (for example Maerzhaeuser, Germany). However results do not differ when the injection is done into one of the cells after the first cleavage. Needles (tip diameter 3-5µm) are pulled on a simple needle puller (for example Narishigi, Japan) from 1.2mm fibre-filled (for rapid filling) glass capillaries. On the evening before injection needles are filled from behind using fine drawn out needles that fit into the inner diameter of the injection needle. DNA at a concentration of 50ug/ml is dissolved in distilled water or Ringer solution and 0.1-0.5% phenol red is added to estimate the 300-500pl injection volume. The needle is connected with thin silicone tubing to a 50ml syringe with which pressure is manually produced. It is important to point out that constant pressure must be provided during the whole injection procedure in order to avoid blockage of the needle.

DNA FACTORS INFLUENCING THE TRANSGENIC APPROACH

DNA Concentration and Survival of Embryos

Because of the rapid cell divisions and cell movements that take place during early medaka and zebrafish embryogenesis, cytoplasmically injected DNA may be quickly distributed and diluted. One might also expect that the injected DNA is rapidly degraded leading to low amounts being available for expression and integration. Injection of large amounts of DNA could minimize these negative effects. However, a high dose of foreign genetic material could also be lethal or interfere with embryogenesis. This effect has been observed in sea urchins and frogs (Rusconi and Schaffner, 1981; McMahon, 1985).

Experiments carried out to determine the highest possible amount of DNA that could be injected into the egg indicated that medaka were more tolerant of DNA than were zebrafish. For example, only 24% of the zebrafish eggs survived for one week following injection of 25 pg DNA (50µg/ml, 0.5nl) whereas 72% of the medaka survived the same time period. When the amount of DNA injected was increased to 50 pg, 16% of the zebrafish and 38% of the medaka survived (Vielkind et al., 1990). The results of these experiments led us to inject 300-500pl of DNA at a concentration of 50 µg/ml (15-25 pg) in all further experiments. This corresponds to 2-3.5 x 10^{10} bp of DNA, or 20-30 times the equivalent of the fish genome (Stuart et al., 1987; Chong and Vielkind, 1989; Vielkind et al., 1990; Stuart et al., 1990; Winkler et al., 1991).

DNA in Linear and Supercoiled Form

The structure of the administered DNA appears to have an effect on its survival, replication, integration and expression in various organisms. For example in the mouse, injected linear DNA appears to be integrated into the host DNA in multiple copies of head-to-tail multimers with high efficiency relative to injected supercoiled forms (Brinster et al., 1985). In sea urchins only linear forms are replicated and concatemerized in head-to-tail, head-to-head, and tail-to-tail fashion (McMahon et al., 1985), while in frogs head-to-tail multimers are formed from supercoiled forms (Rusconi and Schaffner, 1981). In addition, expression of the foreign genes in the recipient appears to be influenced by the structural conformation of the applied DNA. High expression was reported in the frog when the DNA was administered in supercoiled form (Etkin and Balcells, 1985) but in another case efficient transcription was only achieved when the DNA was linearized (Wilson et al., 1986)

Only one detailed study has been carried out to compare the fate of supercoiled versus linear DNA in fish. Chong and Vielkind (1989) followed the fate of the two forms throughout embryogenesis and development in the medaka (Fig. 1). Following

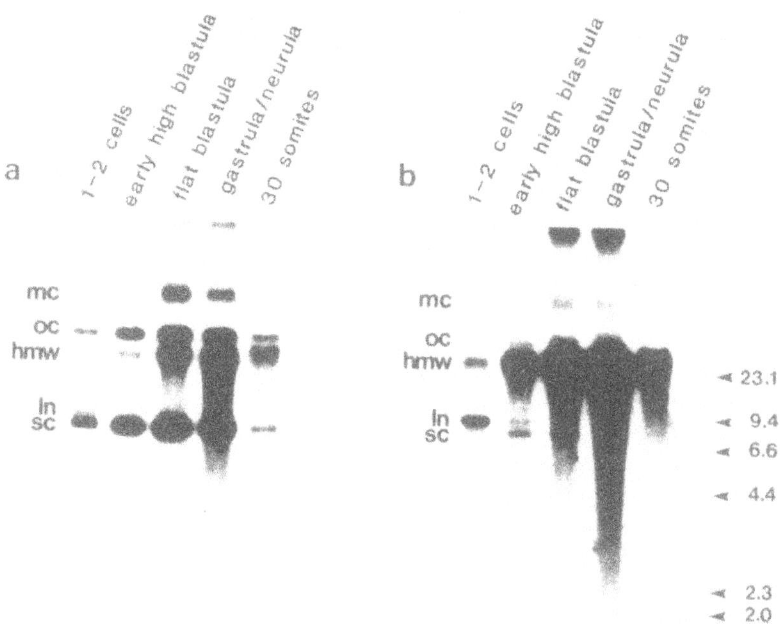

Fig. 1. Fate of pUSVCAT DNA during early medaka embryogenesis after injection as a) supercoiled and b) linear forms into fertilized eggs. Southern blot analyses of DNA equivalent to one embryo out of 10 pooled embryos at various stages of development. hmv = high molecular weight DNA, mc = multimeric plasmid DNA, In = linear DNA, oc = open cirular, and sc = supercoiled plasmid DNA. Numbers = lambda DNA fragments in kb (for details, see Chong and Vielkind, 1989).

injection both linear and supercoiled forms of the injected DNA were rapidly converted into high molecular weight species; linear more rapidly than supercoiled (Fig. 2). The supercoiled plasmid appears to be arranged in head-to-tail multimers in the hatchlings

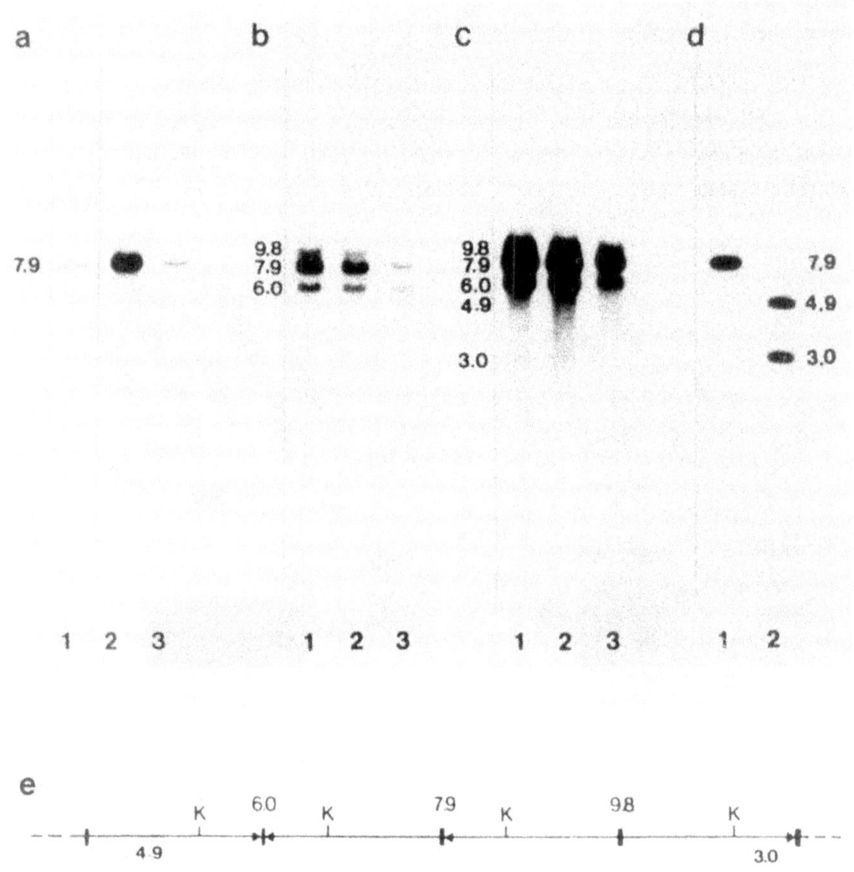

Fig. 2. Nature of high molecular weight pUSVCAT DNA in 3 (lanes 1-3 in a, b, c) individual medaka hatchlings each derived from fertilized eggs injected with a) supercoiled and b) linear pUSVCAT DNA. DNA from individuals was digested with Kpnl and analysed by Southern blot analysis; c) overexposure of blot in b). d) Southern blot analyses of: Kpnl digest of supercoiled (lane 1) and Sall-linearized (lane 2) pUSVCAT DNA; e) diagram of head-to-head, head-to-tail and tail-to-tail multimers derived from linear pUSVCAT ligation products. In pUSVCAT the CAT gene is controlled by the SV-40 early region promoter-enhancer and the promoter-enhancer of the LTR region of the Rous sarcoma virus (for details, see Chong and Vielkind, 1989); K=Kpnl restriction enzyme, numbers = DNA fragments in kb.

indicating homologous recombination of identical linearized molecules (Fig. 2a). In contrast, the linear DNA appears to have been converted into a head-to-tail array but head-to-head and tail-to-tail ligation products also occurred (Fig. 2b, c, e). Most of the injected DNA appeared to be extrachromosomal, however there were indications that some of it was integrated into chromosomal DNA (Chong and Vieklind, 1989; Winkler et al., 1991).

Both forms of DNA began to replicate almost immediately following injection, reaching a peak during the gastrula/neurula stage. This resulted in 10 and 12 fold increases in linear and supercoiled forms respectively. This foreign DNA replication occurred when the embryos were rapidly replicating their own DNA. Stuart et al. (1987) obtained similar results following injection of a linearized construct into zebrafish eggs. Winkler et al. (1991) also observed similar results using a supercoiled plasmid. However these authors (Winkler et al., 1991) also noted a second DNA replication peak one week after injection, a time when rapid growth of the embryo occurs (Fig. 3).

In the study conducted by Chong and Vielkind (1989) on the medaka, both forms of the injected DNA began to show signs of degradation as early as the blastula/gastrula stage. However the foreign DNA was retained by all hatchlings (Fig. 2) and could still be found in 10% of the young and 5% of the adults. These results are similar to those obtained by Winkler et al. (1991) and Stuart et al. (1987).

One important conclusion to be made from these studies on medaka and zebrafish is that both linear and supercoiled forms of the injected DNA can integrate into the germ cells, albeit in a mosaic fashion (Vielkind et al., 1990; Stuart et al., 1990). Stable integration was shown by linkage of the transgene to chromosomal DNA using Southern blot analysis and by Mendelian inheritance in subsequent generations. Inherited expression of the supercoiled transgene has also been observed in offspring from transgenic parents (Vielkind et al., 1990; Stuart et al., 1990). Similar inheritance studies have not been conducted with an appropriate linear sequence. Therefore there is no reason to believe that the supercoiled form is better in this regard than the linear form. For further details and discussions see Stuart et al., 1987; Chong and Vielkind, 1989; Vielkind et al., 1990; Stuart et al., 1990; Winkler et al., 1991.

DNA Application via Phage Particles

One possibile way to introduce foreign DNA into cells or embryos is through the use of modified viral particles. Very efficient gene transfer has been demonstrated in cultured avian and mammalian cells using phage particles as DNA transfer vehicles

(Ishiura et al., 1982; Okayama and Berg, 1985; Vielkind and Vogel, 1989). There are no known fish viruses that can be used for DNA transfer in fish. However experiments demonstrating the feasibility of such an approach have been carried out by Chong and Vielkind (1989) using recombinant phage particles.

Recombinant phage lambda carrying the CAT reporter gene were injected into fertilized medaka eggs. During embryo development the phage DNA behaved in a manner similar to supercoiled and linear forms of DNA with one exception; instead of

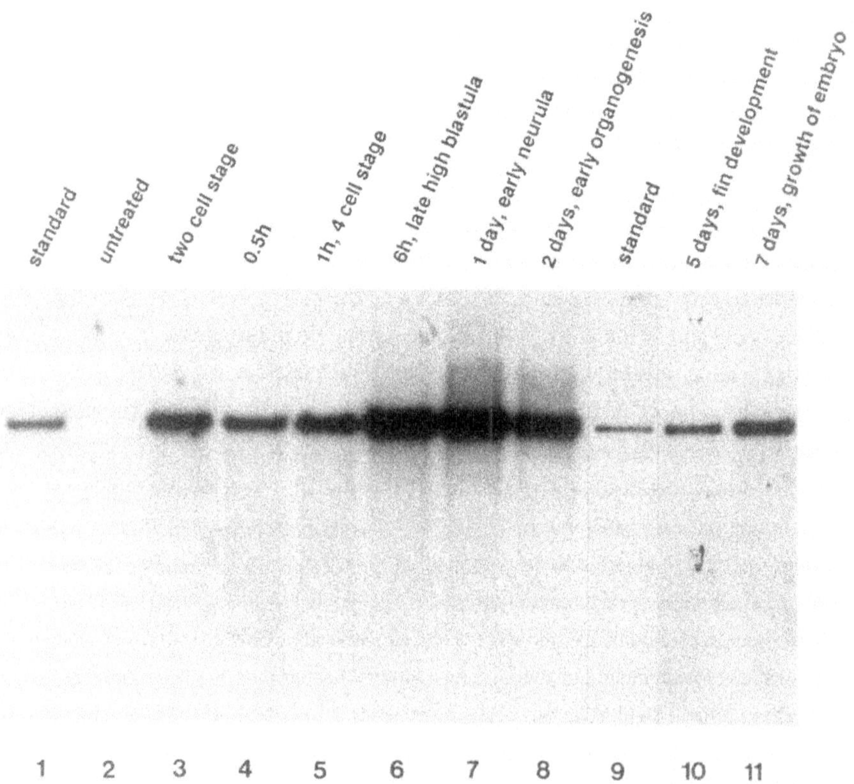

Fig 3. Replication of pCMVTkCAT DNA during early medaka embryogenesis. Southern blot analyses of: XbaI digested DNA from 1.5 equivalents of embryos at various developmental stages after injection with supercoiled pCMVTkCAT DNA at the 1-2 cell stage (lanes 3-11), XbaI digested DNA from uninjected embryos (lane 2) and 40pg XbaI-linearized pCMVTkCAT DNA (lanes 1,9). In pCMVTkCAT the CAT gene is controlled by the human cytomegalovirus enhancer and the thymidine kinase promoter from the *herpes simplex* virus (for details, see Winkler et al., 1991).

declining dramatically after replication, it remained at a constant value (Chong and Vielkind, 1989). As a consequence of the phage DNA persistance a high proportion of young fish expressed the reporter CAT gene. The expression levels in individual embryos were lower than that observed for the linear and supercoiled plasmid forms of the DNA, presumably as a result of the lower amount of phage DNA present compared with the other forms. For technical reasons the number of phage particles that can be injected into medaka eggs is limited. The results of these experiments indicate that phage particles may be excellent vehicles for transfering genes into embryos. Such a method would eliminate the necessity of extracting and purifying the DNA and it would facilitate the transfer of large, up to 20 kb, sequences.

TRANSIENT MEDAKA EXPRESSION SYSTEM

While it is necessary to analyse the true functioning of genes in *in vivo* systems, the functional analysis of genes in a stable transgenic system has several limitations. Firstly, the production of stable transgenics is time consuming, secondly, the integration site and copy number of the integrated transgene may obscure the function of the gene or its control regions, and thirdly, genomic imprinting may alter the expression of the transgene as has been reported in mice (Surani et al., 1988). Transient expression systems may not be affected by these obstacles and thus, could represent alternative systems for gene function analysis. Such systems have been used successfully in the fruit fly (Steller and Pirrotta, 1984), sea urchin (McMahon et al., 1985) and frog (Krone and Heikkila, 1989).

Transient Expression Persists Beyond the Hatching Stage

In order to evaluate whether or not genes cytoplasmically injected into the fertilized medaka eggs can be functionally analysed, the expression of reporter genes encompassed in various plasmid forms or phage and driven by various control regions have been followed during embryogenesis and larval development (Chong and Vielkind, 1989; Vielkind et al., 1990; Winkler et al., 1991). In a basic study the bacterial chloramphenicol acetyltransferase (CAT) gene transcriptionally regulated by the SV-40 origin-early region promoter and the LTR of the Rous sarcoma virus encompassed in the plasmid pUSVCAT (Karisson et al., 1985) was used. Simple and rapid tests are available for measuring the product of this reporter gene. In addition no similar enzyme has been found in eucaryotes thus eliminating interference from endogenous products.

Following microinjection of the various CAT gene constructs the expression of CAT activity was followed throughout embryogenesis, hatching and up to the free swimming fish stage of four weeks post-injection. CAT expression began in the flat blastula stage, peaked in the gastrula/neurula stage and one-week old embryos and decreased but was still detectable in 20-30% of young 4-week old fish. Expression levels appeared to be higher when the supercoiled form was injected. Although CAT expression was lowest when intact phage particles were injected, it persisted in a greater proportion of the four-week old fish than did the supercoiled and linear forms (Chong and Vielkind, 1989).

Promoter-Specific and Tissue-Specific Transient Expression

Promoter-specific expression was studied using the CAT reporter gene driven by three different promoter-enhancers (Winkler et al., 1991). CAT expression was determined at various stages of early development and post-hatch (Fig. 4). All three promoters directed expression of the CAT gene, however there were clear quantitative differences between them at the various stages of development. This suggests that the medaka system responds differentially to the various control regions and therefore could be used as an assay system for gene regulation analysis.

This conclusion is further substantiated by the results obtained from experiments using LacZ as an indicator of tissue specific expression. When the lacZ was controlled by the SV-40 early region promoter, expression was found in neural tissues at 4 and 8 days post injection. When the gene was regulated by the LTR of RSV, expression was found in the integument, myosepta and caudal head mesenchyme. Expression of the LacZ gene was not observed in all cells, but rather exhibited a mosaic distribution throughout each tissue. These observations of tissue-specific expression are similar to those observed in transgenic mice when the same control regions are employed (Brinster et al., 1984; Overbeek et al., 1986).

CHROMOSOMAL INTEGRATION AND INHERITED EXPRESSION OF TRANSGENES IN ZEBRAFISH

It is evident from the above that the medaka embryo is an excellent transient expression system in which to test the functioning of cloned genes. Since most, if not all expression comes from unintegrated genes factors such as chromosomal integration site do not have to be taken into consideration when interpreting the data. However, many studies of gene function *in vivo* require integration of the foreign gene into the host chromosomes of all cells in the body. Such studies need stable germ line integration of the transgene. Zebrafish embryogenesis and development to maturity is

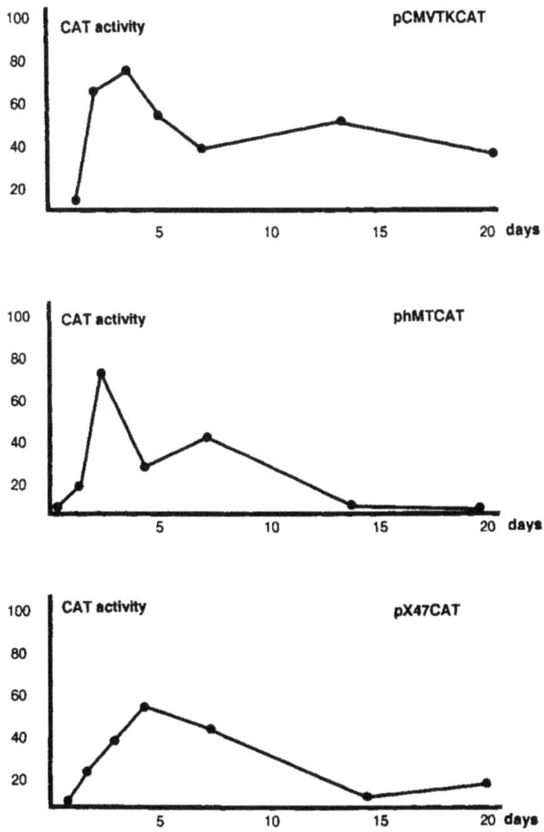

Fig. 4. CAT gene expression pattern during embryonic and larval medaka development. CAT enzyme activity, as assayed by conversion of chloramphenicol to its acetylated forms, in individuals embryos or hatchlings as a result of injection of pCMVTkCAT, phMTCAT and pX47CAT supercoiled plasmid DNA at the 1-2 cell stage. In pCMVTkCAT the cat gene is controlled by the thymidine kinase (Tk) promoter from *Herpes simplex*.In phMTCAT the CAT gene is controlled by the human metallothionein IIA promoter fragment, in pX47CAT it is controlled by a promoter from the platyfish *Xiphophorus maculatus* (for pCMVTkCAT see Fig. 3. For further details, see Winkler et al., 1991).

twice as fast as it is in medaka, therefore zebrafish lend themselves as better a models for the study of a stable transgenic system.

Identification of Stable Transgenic Fish

In the mouse direct pronuclear injection of the foreign genetic material results in almost immediate integration of the transgene into the recipient's DNA in 25% of injected zygotes, 70% of these transgenic mice carry the transgene in all of their cells, including the germ cells, and while the remaining 30% are mosaic (Palmiter and Brinster, 1986). Therefore, a trangenic mouse is easily identified by analysis of DNA extracted from the tip of its tail. In contrast to the mouse, cytoplasmic injections into fish eggs normally result in a mosaic distribution of the injected DNA, not only in specific tissues but also in the fish as a whole. Therefore the presence of a transgene in DNA extracted from a small sample of the fish (fin clips, for example) cannot be taken as evidence for its presence in the germ cells (Stuart et al., 1989).

In order to identify fish that had integrated the transgene (pUSVCAT or pSVeRSVCAT) into the germ cells, Stuart et al., (1990) cross bred DNA treated fish with untreated ones and analysed the offspring (F_1) for the presence of transgene sequences using simple dot blot or transgene expression assays. 5% of the DNA injected fish were able to pass the foreign DNA on to their offspring (Stuart et al., 1990). In three of the four crosses carried out the proportion of offspring inheriting the transgene ranged from 12% to 36% indicating that the parent germ cells were mosiac for the transgene (Stuart et al., 1990). When transgenic offspring (F_1) were crossed with non-transgenic fish 50% of the F_2 offspring inherited the transgene. This result indicates that the transgenic F_1 parents were heterozygous for the foreign DNA which is stably integrated into a single chromosome (Stuart et al., 1987, 1990). This Mendelian inheritance pattern of the transgene argues against its persistence and passage through the germ line as extrachromosomal DNA. Such an inheritance mechanism has been demonstrated in mice and nematodes (Stinchcomb et al., 1985; Rassoulzadegan et al., 1986).

Stable Expression of Transgene in Offspring

In an earlier experiment, Stuart et al., (1987) attempted to produce transgenic zebrafish that were resistant to the antibiotic hygromysin using the plasmid pSV-hygro. However despite the fact that the gene was successfully transferred and passed on to the F_1 generation, there was no evidence for its expression.

Subsequent studies were carried out using the CAT reporter gene driven by the SV-40-LTR promoter-enhancer (pUSVCAT). This gene construct yielded good expression in both the medaka and zebrafish embryos (Chong and Vielkind, 1989; Vielkind, 1990; Stuart et al., 1990). When F_1 offspring resulting from a cross between a transgenic female and a non transgenic male were examined, there was a perfect correlation between the presence of CAT activity and the inheritance of the CAT gene (Fig. 5). Thus it appears that once a fully functional gene is stably integrated into the host chromosomes it will be expressed.

Tissue-Specific and Uneven Expression of Transgene in Offspring

Stuart et al., (1990) studied the tissue distribution of CAT activity in four lines of transgenic F_1 zebrafish. Tissue-specific expression was essentially the same in all four families of fish, with CAT activity being largely confined to fin, skin, heart, gill, skeletal muscle and eyes (Table 1). In three of the transgenic lines the CAT gene was best expressed in the fins and skin, in the fourth line expression was greatest in the heart (Table 1).

The tissue specific activities of the two promoters, RSV-LTR and SV-40, used in tandem in the CAT genes employed in this study have been investigated separately in transgenic mice. In the mouse system the RSV-LTR tends to be expressed in tissues of mesodermal origin. The presence of high levels of CAT expression in heart, muscle and

Table 1. Tissue-specific expression of CAT gene in two (a,b) F_1 individuals from 4 lines (I-IV) of transgenic zebrafish; values are given as CAT activity per gram total protein (for details, see Stuart et al., 1990)

	Ia	Ib	IIa	IIb	IIIa	IIIb	IVa	IVb
Fins	1351	1544	249	109	562	343	663	239
Skin	651	646	120	53	290	99	363	302
Heart	88	69	263	201	69	135	36	*
Muscle	116	111	14	16	15	*	17	86
Gill	69	75	9	15	38	30	145	20
Eyes	45	105	37	27	34	10	12	17
Gut	7	*	8	14	*	*	*	*
Brain	*	*	*	*	*	*	*	*
Liver	*	*	*	*	*	*	*	*
Gonad	*	*	*	*	*	*	*	*

fins of transgenic zebrafish is consistent with this conclusion. However, CAT expression in fins occurred primarily in the epithelial cells, cells that are derived from the ectoderm. It is not known whether expression in these cells is attributable to specificity directed by the RSV-LTR, to the SV-40 or to some synergistic effect of all, or some of the sequences used in this CAT gene construct.

Attempts were made to determine the cell types responsible for CAT expression using a monoclonal antibody to CAT. Immunoreactivity was readily observed in tissues exhibiting high expression levels, but difficult to detect in other tissues. This suggests

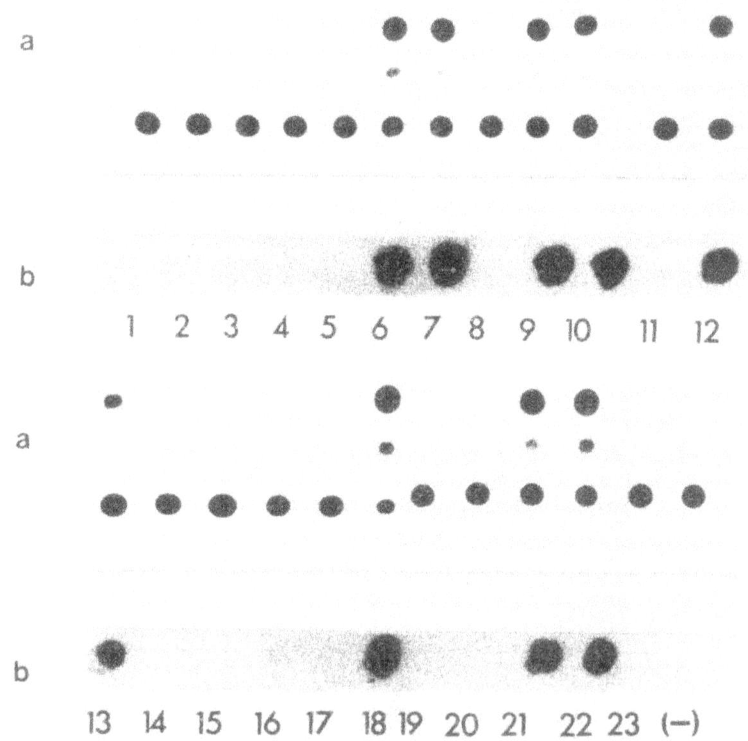

Fig. 5. Mosaic germ line integration of transgene. CAT gene inheritance in individual F_1 resulting from a cross of a positive founder with an uninjected fish; 9 out of 23 F_1 showed perfect correlation between a) CAT gene expression as assayed by CAT enzyme activity assay and b) presence of CAT gene as analysed by dot-blot DNA analysis. For further detail, see Stuart et al., 1990.

that the immunohistochemical procedure was less sensitive than the enzymatic CAT assay. One surprising result from this study was the observation that expression levels of individual cells within fin tissue varied widely, ranging from undetectable to high. It is possible that this varigated expression pattern is attributable to the fact that these F_1 transgenic fish were heterozygous for the CAT gene. However a similar immunohistochemical study of homozygous F_2 transgenics exhibited a similar expression pattern. Potential reasons for varigated expression have been discussed in detail by Stuart et al., (1990) and will not be reiterated have except to point to the possibility that it could reflect a normal form of gene regulation such as has been observed by others (Chelly et al., 1989).

All of these experiments clearly demonstrate that stable, tissue dependent expression of a transgene can be achieved using the zebrafish as a model system.

FINAL REMARKS

For the researcher interested in deciphering molecular mechanisms governing development as well as for the applied scientist interested in improving food production the transgenic system has become an essential tool. It not only facilitates fundamental studies of gene structure and function it also provides the means of altering the genetic makeup and thus the phenotype of important species.

The experiments described in this manuscript underline the value of the medaka and zebrafish as model transgenic systems. The experiments carried out to date only establish the basic information. Future research will include the use of fish promoter-enhancers and fish genes. Detailed comparisons of stable inheritance and expression of linear versus supercoiled constructs, with and without the plasmid, or other vector sequences have yet to be performed. In addition further work needs to be done to find the best way to transfer foreign genes to fish. Clearly considerable research has yet to be carried out before the transgenic methodology in fish becomes as routine as in mice.

REFERENCES

Allen, N.D., Cran, D.G., Barton, S.C., Hettie, S., Reik, W. and Surani, M.A. (1988) Transgenes as probes for active chromosomal domains in mouse development. Nature 333: 852-855.

Brinster, R.L., Chen, H.Y., Messing, A., van Dyke, T., Levine, A.J. and Palmiter, R.D. (1984) Transgenic mice harboring SV40 T antigen genes develop characteristic brain tumors. Cell 37: 367-379.

Brinster, R.L., Chen, H.Y., Trumbauer, M.E., Yagle, M.K. and Palmiter, R.D. (1985) Factors affecting the efficiency of introducing foreign DNA into mice by microinjecting eggs. Proc. Natl. Acad. Sci. USA 82: 4438-4442.

Brown, D.D. and Gurdon, J.B. (1977) High-fidelity transcription of 5S DNA injected into Xenopus oocytes. Proc.Natl. Acad. Sci. USA 74: 2064-2068.

Chelly, J., Concordet, J-P., Kaplan, J-C. and Kahn, A. (1989) Illegitimate transcription: transcription of any gene in any cell type. Proc. Natl. Acad. Sci. USA 86: 2617-2621.

Chong, S.S.C. and Vielkind, J.R. (1989) Expression and fate of CAT reporter gene microinjected into fertilized medaka (*Oryzias latipes*) eggs in the form of plasmid DNA, recombinant phage particles and its DNA. Theor. Appl. Genet. 78: 369-380.

Chourrout, D., Guyomard, R. and Houdebine, L.M., (1986) High efficiency gene transfer in rainbow trout (*Salmo gairdneri* Rich) by microinjection into egg cytoplasm. Aquaculture 51: 143-150.

Etkin, L.D. and Balcells, S. (1985) Transformed Xenopus embryos as a transient expression system to analyze gene expression at the midblastula transition. Dev Biol 108: 173-178.

Fletcher, G.L., Shears, M.A., King, M.J., Davies, P.L. and Hew, C.L. (1988) Evidence for antifreeze protein gene transfer in Atlantic salmon (*Salmo salar*). Can. J. Fish. Aquat. Sci. 45: 352-357.

Fletcher, G. L. and Davies, P. L. (1991) Transgenic fish for aquaculture. In: Genetic engineering, principles and methods. J. K. Setlow (Ed.): 331-370.

Gordon, J.W., Scangos, G.A., Plotkin, D.J., Barbosa, J.A. and Ruddle, F.H. (1980) Genetic transformation of mouse embryos by microinjection of purified DNA. Proc. Natl. Acad. Sci. USA L7: 7380-7384.

Gritz, L. and Davies, J., (1983) Plasmid-encoded hygromycin B resistance: the sequence of hygromycin B phosphotransferase gene and its expression in *E. coli* and *S. cerevisiae*. Gene 25: 179-188 .

Hammer, R.E., Pyrsel, V.G., Rexroad, C.E. Jr., Wall, R.J., Bolt, D.J., Ebert, K.M., Palmiter, R.D. and Brinster, R.L. (1985) Production of transgenic rabbits, sheep and pigs by microinjection. Nature 315: 680-683 .

Hyodo-Taguchi, Y. and Egami, N. (1985) Establishment of inbred strains of the medaka *Oryzias latipes* and the usefulness of the strains for biomedical research. Zool. Sci. 2: 305-316.

Inoue, K., Ozato, K., Kondoh, H., Iwamatsu, T., Wakamatsu, Y., Fujita, T. and Okada, T.S. (1989) Stage-dependent expression of the chicken α-crystallin gene in transgenic fish embryos. Cell Differ. Develop. 7: 57-68 .

Inoue, K., Yamashita, S., Hata, J-i., Kabano, S., Asada, S., Nagahisa, E. and Fujita, T. (1990) Electroporation as a new technique for producing transgenic fish. Cell Differ. Develop. 9: 123-128.

Ishiura, M., Hirose, S., Uchida, T., Hamada, Y., Suzuki, Y. and Okada, Y. (1982) Phage particle mediated gene transfer to cultured mammalian cells. Mol. Cell. Biol. 2: 607-616.

Jaenisch, R. and Mintz, B. (1974) Simian virus 40 DNA sequences in DNA of healthy adult mice derived from preimplantation blastocysts injected with viral DNA. Proc. Natl. Acad. Sci. USA 71: 1250-1254.

Karisson, S., Humphries, R.K., Gluzman, Y. and Nienhuis, A.W. (1985) Transfer of genes into hematopoietic cells using recombinant DNA viruses. Proc. Natl. Acad. Sci. USA 82: 158-162 .

Kimble, J., Hodgkin, J., Smith, T. and Smith, J. (1982) Suppression of an amber mutation by microinjection of suppressor TRNA in *C. elegans*. Nature 299: 456-458 .

Krone, P.H. and Heikkila, J.J. (1989) Expression of microinjected hsp7O/CAT and hsp3O/CAT chimeric genes in developing *Xenopus laevis* embryos. Development 106: 271-181.

Laale, H.W. (1977) The biology and use of zebrafish. A literature review in fisheries research. J. Fish Biology 10: 121-173 .

Maclean, N, Penman, D. and Zhu, Z. (1987) Introduction of novel genes into fish. Bio/Technology 5: 257-261.

McMahon, A.P., Flytzanis, C.N., Houg-Evans, B.R., Katula, K.S., Britten, R.J. and Davidson, E.H. (1985) Introduction of cloned DNA into sea urchin egg cytoplasm: replication and persistence during embryogenesis. Dev. Biol. 108: 420-430.

Matsui, K. (1949) Illustration of the normal course of development in the *Oryzias latipes* (in Japanese). Jpn. J. Exp. Morphol. 5: 33-42.

Okayama, H. and Berg, P., 1985) Bacteriophage lambda vector for transducing a cDNA clone library into mammalian cells. Mol. Cell. Biol. 5: 1136-1142.

Overbeek, P.A., Lai, S., Van Quil, K.R. and Westphal, H. (1986) Tissue-specific expression in transgenic mice of a fused gene containing RSV terminal sequences. Science 231: 1574-1577.

Ozato, K., Kondoh, H., Inohara, H., Iwamatsu, T., Wakamatsu, Y. and Okada, T.S. (1986) Production of transgenic fish: introduction and expression of chicken α-crystallin gene in medaka embryos. Cell Differ. 19: 237-244.

Ozato, K., Inoue, K. and Wakamatsu, Y. (1989) Transgenic fish: biological and technical problems. Zool. Sci. 6: 445-457.

Ozato, K., Inoue, K. and Wakamatsu, Y. (1992) Gene transfer and expression in medaka embryos. In: Transgenic Fish. C. L. Hew and G. Fletcher (eds.) World Scientific Publishing Co, Singapore, pp. 27-43.

Palmiter, R.D. and Brinster, R.L. (1986) Germ-line transformation of mice. Ann. Rev. Genet. 20: 465-499.

Rassoulzadegan, M., Leopold, P., Vailly, J. and Cuzin, F. (1986) Germ-line transmission of autonomous genetic elements in transgenic mouse strains. Cell 46: 513-519 .

Rubin, G.M. and Spradling, A.C. (1982) Genetic tranformation of *Drosophila* with transposable element vectors. Science 218: 348-353.

Rusconi, S. and Schaffner, W. (1981) Transformation of frog embryos with a rabbit B-globin gene. Proc. Natl. Acad. Sci. USA 78: 5051-5055.

Schnieke, A., Harbers, K. and Jaenisch, R., (1988) Embryonic lethal mutation in mice induced by retrovirus insertion into the alphal (I) collagen gene. Nature 304: 315-320.

Sinn, E., Muller, W., Pattengale, P., Tepler, I., Wallace, R. and Leder, P. (1987) Coexpression of MMTV/v-Ha-ras and MMTV/c-myc genes in transgenic mice: synergystic action of oncogenes in vivo. Cell 49: 465-475.

Spradling, A.C. and Rubin, G.M. (1982) Transposition of cloned P-elements into Drosophila germ line chromosomes. Science 218: 341-347.

Steller, H. and Pirrotta, V. (1984) Regulated expression of genes injected into early Drosophila embryos. EMBO J. 3: 165-173 .

Stinchcomb, D.T., Shaw, J.E., Carr, S.H. and Hirsh, D. (1985) Extrachromosomal DNA transformation of *Caenorhabditis elegans*. Mol. Cell. Biol. 5: 3484-3496.

Streisinger, G., Walker, C., Dower, N., Knauber, D. and Singer, F. (1981) Production of clones of homozygous diploid zebra fish (*Brachydanio rerio*). Nature 291: 293-296.

Stuart, G.W., McMurray, J.V. and Westerfield, M., (1987) Replication, integration and stable germline transmission of foreign sequences injected into early zebrafish embryos. Development 103: 403-412.

Stuart, G.W., Vielkind, J.R., McMurray, J.V. and Westerfield, M. (1990) Germ-line transmission and variegated expression of a foreign gene in transgenic zebrafish. Development 109: 577-584.

Surani, M.A., Reik, W. and Allen, N.D. (1988) Transgenes as molecular probes for genomic imprinting. Trends Genet. 4: 59-62.

Vielkind, J.R. and Vogel, K.S. (1989) Gene transfer and expression studies in cultured avian neural crest cells differentiating into melanocytes. Pigm. Cell Res. 2: 44-52.

Vielkind, J.R., Chong, S.S.C., Stuart, G.W. and Sadaghiani, B. (1990) Transgenic medaka and zebrafish systems: Transient and inherited expression of a recombinant CAT reporter gene microinjected into fertilized eggs, In: New Trends in Ichthyology, Schroeder, H.J. and Schartl, M. (eds.), Paul Parey, Munich.

Wilson, C., Cross, G.S. and Woodland, H.R. (1986) Tissue-specific expression of actin genes injected into *Xenopus* embryos. Cell 47: 589-599.

Winkler, C., Vielkind, J.R. and Schartl, M. (1991) Transient expression of foreign DNA during embryonic and larval development of the medaka fish (*Oryzias latipes*). Mol. Gen. Genet.(in press).

Yamamoto, T. (1967) Medaka. In: Methods in Developmental Biology, Wilt, F.M. and Wessels, N.K.(eds.), Growell, NewYork, pp. 101-111.

CHAPTER 6

Generation of Fast Growing Transgenic Fish: Methods and Mechanisms

Zuoyan Zhu

ABSTRACT

Fish are an excellent system for the study of transgenesis. By the technique of microinjection, and recently by electroporation, foreign genes, e.g. PMhGH fusion gene, are easily introduced into fish fertilized eggs after chorion removal. About 50% of the microinjected and 10% of the electroporated fertilized eggs developed into transgenic fish. As soon as the foreign gene was introduced into the embryos, the linear DNA was immediately switched into different physical forms. Onset of foreign gene replication occurred between the blastula and the gastrula stages. Integration of polymers of the foreign gene mainly occurred at late stages of embryogenesis. This eventually led to generation of mosaic transgenic fish and segregation into offspring by germ line transmission. Most copies of the transgene maintained their integrity in the host chromosomes of both P_1 and F_1 generations. Transcripts of the hGH gene in linear form, but not of the circular form, were generated starting from early stages of embryogenesis. The final expression product, human growth hormone, was produced and stimulated growth rates by 2-4 times in some juveniles. However, both the MT-1 gene promoter and the hGH structural gene are not suitable for the purpose of fish breeding. For this reason, we have made a new construct of pCAgGH, i.e. a common carp ß-actin gene promoter and presumptive control region driving a grass carp growth hormone gene in pUC118. In combination with other techniques, such as cell culture, nuclear transplantation, gynogenesis or androgenesis, induced polyploidy and sex reversal, transgenic animals with ecological acceptability will hopefully be a reality in fish.

INTRODUCTION

Fish culture is one of the earliest activities of man. The first record of pond culture in China can be dated back 2500 years to the *Handbook of Fish Culture*, in which the author, Fan Li, described the domestication and cultivation of common carp in ponds. Advancement in the techniques of fish culture have increased with an expanding human population. However, the productivity of many culture fishes is restricted by the natural physiochemical performances of the species concerned. For instance, rainbow trout now named *Oncorhynchus mykiss* and salmon cannot be stocked in warm water while tilapias (e.g. *Oreochromis niloticus*) can only be raised in warm water. In addition, various fish species are also limited in their growth by viral infection. Reo-type virus infection in grass carp (*Ctenopharyngodon idellus*) can account for 50-80% of juvenile mortality. Thus, in order to improve breeding qualities of important farmed fish species, it is necessary to develop new techniques, other than those which rely on classical genetic selection and cross hybridization methods.

During the last two decades, a few important techniques have been developed which have revolutionized methods in fish biology and breeding. These techniques, generally referred to as genetic manipulation include induced polyploidy (Thorgaard, 1986; Chourrout and Quillet, 1982), monosexual reproduction (gynogenesis and androgenesis) (Purdon, 1984), induced maturation and ovulation (Wu and Chung, 1964, Donaldson and Hunter, 1983), sexual transformation (Bye and Lincoln, 1986), "nuclear-cytoplasm hybridization" (nuclear transplantation) (Tung et al., 1963), and gene transfer (Zhu et al., 1985). In this paper, we will focus on the methods and mechanisms of generating transgenic fish.

The ability to conduct transgenic projects is based on two major advances in molecular and developmental biology. The first advancement was the development of molecular cloning techniques which enabled the isolation and cloning of a single gene which codes for a unique protein within the organism. The second advancement was the technique of embryo micro-manipulation which enabled "foreign gene(s)" to be introduced into an organism's genome and allowed "foreign protein(s)" to be expressed within that organism. Early successful transgenic experiments were demonstrated in mammals (Gordon and Ruddle, 1981; Wagner et al., 1981a,b; Constantini and Lacy, 1981), and in *Xenopus laevis* (Bending, 1981; Rusconi and Schaffner, 1981; Etkin et al., 1984). One example of gene transfer was the work done by Palmiter et al. (1982) in the mouse. By microinjection, they introduced into the male pronucleus of mouse fertilized eggs a fusion gene construct of the mouse metallothionein-I promoter driving a rat growth hormone gene. As a result of expression of the foreign gene, some of the transgenic mice showed a doubling in size at maturation. Based on the same strategy, we succeeded in producing the first batch of transgenic fish in 1984 and subsequently explored the mechanisms controlling the foreign gene integration, expression, the biological function and the pattern of inheritance by germ line transmission (Zhu et al.,

1986, 1989; Maclean et al., 1987). In recent years, more than ten laboratories throughout the world have been successful in generating transgenic fish in a variety of species using different foreign gene constructs (Chourrout et al., 1986; Ozato et al., 1986; Maclean et al., 1987; Dunham et al., 1987; Fletcher et al., 1988; Stuart et al., 1988; McEvoy et al., 1988; Brem et al., 1988; Rokkones et al., 1989; Yoon et al., 1990; Zhang et al., 1990).

Preparation of Fertilized Fish Eggs

A variety of species of freshwater fish have been used in our laboratory for the study of transgenesis. These are: goldfish or crucian carp (*Carassius auratus auratus*), loach (*Misgurnus anguillicaudatua*), common carp (*Cyprinus carpio*) including varieties of mirror carp and red carp, and silver crucian carp (*Carassius auratus gibelio*). Goldfish, common carp and silver crucian carp are spring spawners, while loach spawns from spring to fall. During early spring mornings, when the water temperature rises to 17-18°C, mature male and female fish chase after each other around water-weeds. This behavior pattern eventually stimulates females to release their eggs and males to excrete their milt. Once the eggs have been covered with milt, fertilization immediately follows. The eggs of all four species share similar characteristics: 80-150 µm in diameter; transparent or translucent animal pole; semi-viscous chorion which allows eggs to adhere to surfaces of water weeds or other objects. Another unique characteristic is that the chorion of these eggs can be digested with the serine protease, trypsin, without harming normal egg development. This result was revealed earlier in a study with another freshwater fish species, *Clarias fuscus* (Zhu, 1982).

In order to control the time course of gene introduction into developing fish embryos, artificial fertilization is required. After an overnight hormonal induction (800-1000 units of chorionic gonadotropin/ kg body weight or 5 mg of common carp pituitary homogenate/ kg body weight), the mature eggs and milt were striped from parental fish. About two hundred eggs were collected in a Petri dish and immediately mixed with a tiny drop of milt. The eggs were then allowed to stand for 5 minutes while fertilization took place. To remove the chorion, eggs were rinsed gently twice with water and then soaked in 0.25% trypsin solution (in 0.7% NaCl or Holtfreter's solution) with occasionally tilting. The process of chorion digestion took about 5 minutes and could be monitored under a stereomicroscope (10-20x) (Fig. 1: A, B, C). When the chorion became thinner and more transparent, but not completely digested off, the trypsin solution was replaced with fresh Holtfreter's solution. The remaining enzyme on the chorion completed the course of digestion without any damage to the embryos. About 100-200 naked eggs were collected by aspiration with a wide mouth glass pipet, into an

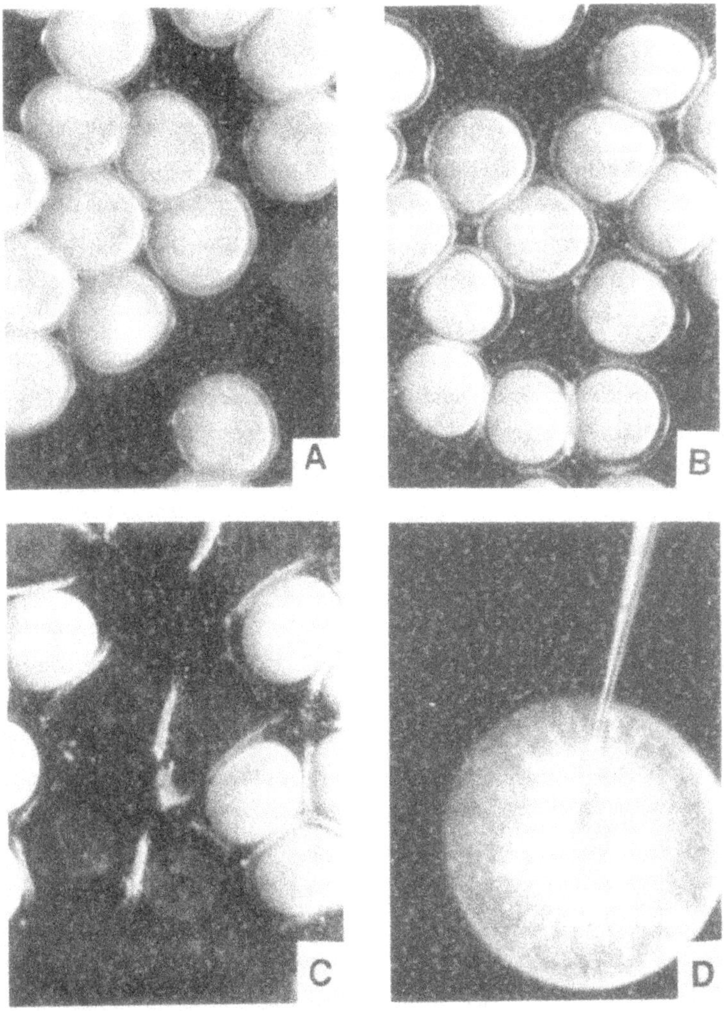

Fig. 1. The trypsin digestion of crucian carp egg chorion. A, the newly fertilized eggs attached to the bottom of a petri dish; B, three minutes after 0.25% trypsin treatment (20°C), the chorion of the eggs is thinner and more transparent. The trypsin solution is then replaced by Holtfreter's solution; C, 2-3 minutes after the solution replacement, the chorion is digested off; D, 1-2 nl of DNA solution (10^5-10^6 copies of foreign gene) is released around the area of the egg's pronucleus by microinjection (A, B and C: 20x; D: 70x).

agar layered petri dish filled with Holtfreter's solution. At this time, the eggs were ready for micromanipulation.

Preparation of Foreign Gene(s)

For the choice of a "foreign gene" to be introduced in transgenic fish, the following criteria were taken into account. The gene should be: 1) specifically detectable when transferred into the host, i.e. little or no sequence homologey between the foreign gene and the host genome; 2) easy to assay for gene expression, if any, at the transcriptional and the translational levels; 3) regulatable in terms of the control of gene expression; and 4) one in which the expressed product exerts some biological function in the host so that the resulting transgenic fish could be selected by physiological and/or morphological characteristics.

In our work, the most important precondition was the availability of cloned genes. One possibility was the human insulin gene (Zhu, Z. and Fritsch, E. F., 1982, unpublished data), but the effect of insulin on fish could not be predicted. Fortunately, we received the recombinant plasmid DNA construct pBPVMG from Dr. D. Hamer. The pBPVMG is a pBR322 recombinant clone. The insert fragment of DNA contains 9.4 kilobase pairs (kb) of nucleotide including 69% bovine papilloma virus (BPV) transforming sequence followed by a MhGH fusion gene, i.e. the mouse metallothionein-1 gene promoter and presumptive control region driving a human growth hormone structural gene (with the first intron). It had earlier been demonstrated that mouse cells transformed with this recombinant construct synthesized a metallothionein-growth hormone hybrid mRNA and produced mature human growth hormone. The synthesis of hybrid mRNA was regulated by cadmium administration (Pavlakis and Hamer, 1983). In addition, there was strong evidence suggesting that fish could respond to mammalian growth hormone treatment (Pickford, 1954). Moreover, there was no cross reactivity between growth hormone of human and fish (at least in the four species mentioned above) by radioimmunoprecipitation assay and no cross reactivity between the growth hormone genes by Southern hybridization (Zhu et al., 1985; Zhu et al., 1989). Therefore, pBPVMG met all of the criteria thought necessary for the selection of a foreign gene for the study of transgenics.

The microinjection technique made it possible to introduce DNA solution directly into specific parts of a fish egg or embryo, therefore the papilloma virus transforming sequence was spliced away from pBPVMG, generating the pMhGH construct used in most of our experiments. The human insulin gene was also linked to pMhGH, generating the recombinant plasmid pMhGI, in which the human growth hormone gene was driven by the mouse metallothionein-1 gene promoter and the human insulin gene was driven by its own promoter (Fig. 2).

Recombinant plasmid DNA of pBPVMG, pMhGH and pMhGl were prepared according to Maniatis et al. (1982). With appropriate restriction enzyme digestion, linearized DNA fragments containing hGH or hGl gene(s) with or without the plasmid sequence were separated by agarose gel electrophoresis and purified by phenol/chloroform extraction. After ethanol precipitation and extensive washing with 70% ethanol, the DNA was dissolved in ST buffer (88 mM NaCl and 10 mM Tris, pH 7.5) at an injection concentration of 25-50 ng/μl.

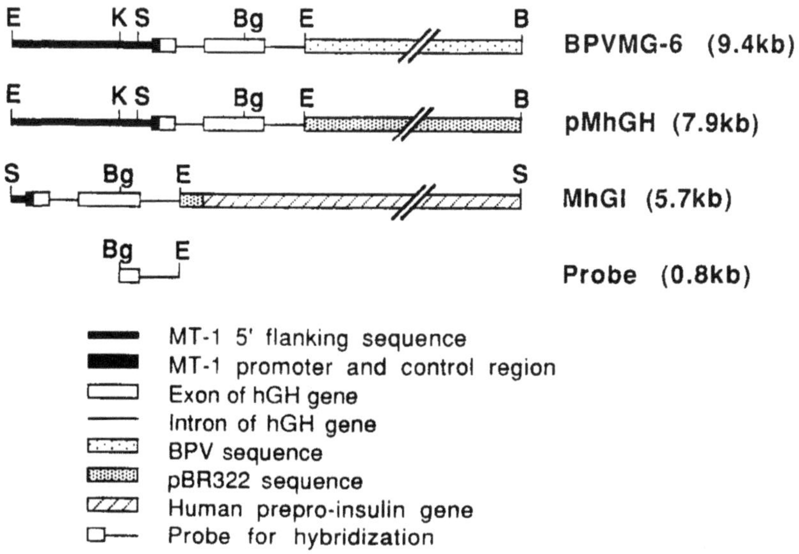

Fig. 2. Constructs of the hGH gene used for generating transgenic fishes. BPVMG-6 originally came from Dr. Dean H. Hamer's Laboratory, NIH, USA. The MT-1 fragment is the promoter and control region of the mouse metallothionein-I gene, which contains 1900 bp of the 5' flanking sequence are 68 bp of the 5' untranslated sequence. The hGH fragment is 1500 bp of the human growth hormone gene which lacks all of the 5' flanking sequences and the last three of the four introns found in the chromosomal gene. *In the MhGl, a 3500 bp human preproinsulin gene sequence is fused to the 3' end* of the hGH fragment along with a short piece of the pBR322 sequence. (B, Bam HI; Bg, Bgl 1; E, Eco RI; K, Kpn 1; S, Sac 1)

INTRODUCTION OF THE FOREIGN GENE INTO FISH EGGS

The objective of our study is the generation of a transgenic strain, rather than a transgenic individual. In an ideal transgenic fish, foreign genes should spread into all cells. The copy numbers and sites of foreign gene integration in the chromosomes of all cells must be as uniform as possible. To achieve this goal, the foreign gene was directly delivered into the earliest embryonic stage possible in order to facilitate the early integration into the host zygotic chromosomes. Fertilized eggs of the four species of fish mentioned above develop quickly. For example, the eggs spawned by the common carp are at the metaphase stage of second meiosis when the sperm enters the micropyle (Ge and Jiang, 1985). About 12 minutes later, the second polar body forms and is ready to be extruded. The egg chorion can be removed with trypsin treatment at this time. Under a stereomicroscope, the second polar body is a tiny "glittering ball" (about 8 μm in diameter) located on the central surface of the animal pole. Even though the bulk of yolk prevents the egg pronucleus from being visible, one can estimate its location from the tiny "glitter". Both egg and sperm pronuclei are located just beneath the second polar body. About 26-30 minutes after fertilization the membrane of the pronuclei disappears, the two pronuclei fuse with each other, and a zygotic nucleus is formed. In the following 20-30 minutes, the zygotic nuclear membrane disappears and the chromosomes move toward both ends of the spindle. Finally, the blastomere appears on top of the yolk sac. The best time to load a foreign gene into the egg and facilitate the gene mixing with the zygotic nucleus or chromosomal mass is 10-30 minutes after fertilization, during which time the second polar body is distinguishable as a location marker. For experimental purposes, we carry out microinjection up until first cleavage.

The microinjection instrument used in our laboratory was constructed in the Institute Workshop, and was similar to those used for nuclear transplantation in the frog (Briggs and King, 1952; Gurdon, 1962) and fish (Tung et al., 1963) . Glass micro-needles, with a tip inner diameter of 3-4 μm, were drawn with a needle drawing apparatus (Leitz). About 2-3 μl of foreign gene solution were usually loaded into each micro-needle. The needle containing the DNA sample was connected to a 50 ml syringe with a piece of plastic tube filled with liquid paraffin. Dechorionated eggs were lined up with the germinal disc facing upwards in the agar layered petri dish. The arm of the micromanipulator was carefully adjusted to direct the tip of micro-needle above the center of the germinal disc, where the second polar body should be located. The needle was lowered slowly until it touched the surface of the embryo. If the tip of the needle was smooth enough, DNA solution automatically flowed out, otherwise light pressure on the syringe was required. The needle tip was dipped down about one third of the thickness of the germinal disc and the DNA solution was released around the area of the pronuclei or zygotic nucleus (Fig. 1: D). Since there was a constant flow of DNA solution through the glass microneedle, the foreign gene copy number received by each egg was controlled by the timing of the needle-dipping interval. We routinely inject 10^5-10^6 copies of foreign gene, in about 2 nl, per egg. A well trained person can inject 100-150 eggs per batch before the first cleavage occurs (within 30-40 minutes).

Recently, we have succeeded in transferring foreign genes into fish by the technique of electroporation (Xie et al., 1989). The electroporation system was designed and made in our laboratory (Fig. 3). The optimum paremeter settings for transfer of the pMhGH gene into dechorionated eggs were determined (Table 1). A setting of 250 V/ 22 μF/ 20Ω /1 mS resulted in about 10% of the treated loach eggs developing into transgenic individuals with 1-77 copies of foreign gene per genome.

TRANSGENE BEHAVIOUR DURING HOST EMBRYOGENESIS

After microinjection, the fertilized eggs were carefully transferred into a large Petri dish containing fresh Holtfreter's solution. Stocking density was kept low. Abnormal embryos were counted and removed from time to time. A number of uninjected control eggs from the same batch were raised in the same way. To facilitate further study, embryogenesis was divided into the following stages: 1 - 2 cell stage,

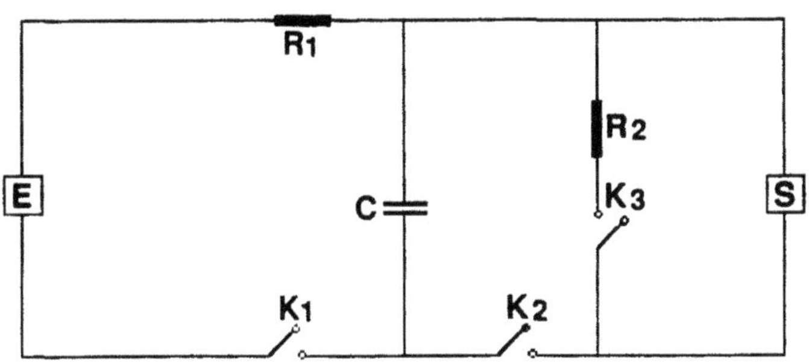

Fig. 3. Diagram of an electroporation system for transgenic fish. E, power supply (D. C.); C, capacitor; R, resistance; S, sample chamber; K, key

Table 1. The influence of electroporation on the development of fertilized loach eggs (E=250 V).

	Parameter settings			Development stages		
Experiment no.	C (μF)	R2 (Ω)	P.T. (mS)	M	G_2	F
1	0.047	∞	3.3	++	++	++
2	0.07	∞	5.1	++	++	++
3	22.0	20	5.1	++	++	++
4	22.0	180	8.7	+	-	--
5	22.0	1000	48.4	-	--	--
6	150.0	20	6.6	+	-	--
7	150.0	180	59.4	-	--	--
8	150.0	500	165.0	-	--	--
9	150.0	∞	$1.1x10^5$	*		

(P.T., pulse time; M, multicellular stage; G_2, late gastrula stage; F, feeding stage; *, fertilized eggs collapsed immediately)

multi-cell (64-128 cells) stage, blastula stage, early and late gastrula stages, neuronal stage, tail bar stage or muscular reaction stage, heart beating stage, pigment formation stage, and feeding stage.

Once microinjection or electroporation had been done, the first question to be considered was whether and how the foreign genes "survived". Initially, a simple and straightforward method, "embryo hybridization", was developed in our laboratory. The samples of gastrula embryo (9-10 hours after fertilization) were removed from the culture dish and washed five times with 0.7% NaCl. The cell mass of the embryo plate was peeled from the yolk sac with a glass knife (a piece of round spun glass) and transferred into calcium ion free Holtfreter's solution for 10 minutes to allow cell dissociation. The dissociated cells were then transferred onto a nitrocellulose filter and subjected to cell lysis, denaturation with 0.5N NaOH, 1.5M NaCl, and neutralization with 1.5M NaCl, 0.5M Tris.HCI (pH 7.5). The nitrocellulose filter was baked at 80° C in a vacuum oven for 2 hours and then hybridized to the ^{32}P-labeled foreign gene, according to conditions described by Grunstein et al. (1975). Interestingly, the first batch of embryo hybridizations indicated that all individual host gastrulae were strongly positive for the ^{32}p-labeled foreign gene probe (all of the control samples were negative) (Zhu et al., 1989). This result clearly revealed that foreign genes, introduced at the beginning of fish embryogenesis, not only existed but also amplified through the gastrula stage.

The next questions were whether and when the foreign gene integrated into the host genome. At each specified development stage, random samples of 20-30 embryos were pooled and total DNA samples were isolated by the procedure of Rusconi et al. (1981). DNA equivalents of 2 embryos, at progressive development stages, were dropped onto a nitrocellulose filter for dot blot hybridization with the ^{32}P-labeled foreign gene probe. DNA samples from 4 embryos at the same stage were separated on a 0.8% agarose gel by electrophoresis, Southern blotted onto a nitrocellulose filter, and hybridized with the same radioactive probe. Dot blot hybridization revealed that replication of the foreign gene started immediately once introduced into fertilized fish eggs, and the strongest signal of replication appeared during late blastula to early neurula stages. After the neurula stage, the total amount of DNA hybridized to the ^{32}P-labeled probe gradually decreased but remained detectable at the feeding stage. The result of Southern blot hybridization suggested a dynamic process of foreign gene replication, degradation, and possibly integration during embryogenesis. Within a few minutes transfer of the linear form of the foreign DNA was converted into different physical forms; circular, dimers, and concatamers. Dimers or concatamers were the prefered form of foreign gene replication when the embryo developed from the multi-cell to the late gastrula stage. Starting from the neurula stage, the molecular weight of detectable foreign DNA increased while its total amount decreased suggesting that the foreign gene formed multimers and/or increased in copy number in the host. The unintegrated circular, monomer and dimer forms of the DNA were degraded after the muscular reaction stage (Fig. 4).

To identify integration of a foreign gene into the host genome, total DNA samples were extracted from fry or from different tissues (muscle, erythrocytes, liver, and testis) of adult fish. Dot blot hybridization of the total DNA samples with a ^{32}P-labeled foreign gene probe revealed the presence of about 50% positive individuals in the treated population (Table 2). Southern blot hybridization indicated that all of the foreign genes were associated with cellular DNA. The positive individuals were termed "transgenic fish" and the foreign genes integrated in their chromosomes were accordingly referred to as "transgenes".

Transgenic red crucian carp with the transgene MhGI/ Sac I were generated and examined for transgene integration (Fig. 5). MhGI/ Sac I is a 5.7 kb linear recombinant DNA fragment containing a human growth hormone gene, a human insulin gene, and a mouse MT-1 gene control region. Total cellular DNA was isolated from a transgenic red crucian carp fingerling and divided into three aliquots (10 μg each). Two

Table 2. The integration ratio of MhGH gene in fishes.

Fish species	Number of fish tested	Number of positive fish	Integration ratio (%)
Loach	10	4	40.0
Crucian carp	64	33	51.6
Red crucian carp	66	28	42.4
Silver crucian carp	23	16	69.6
Mirror carp	89	46	51.7
Red carp	8	7	87.5
Loach*	96	10	10.4

(* by electroporation, others by microinjection)

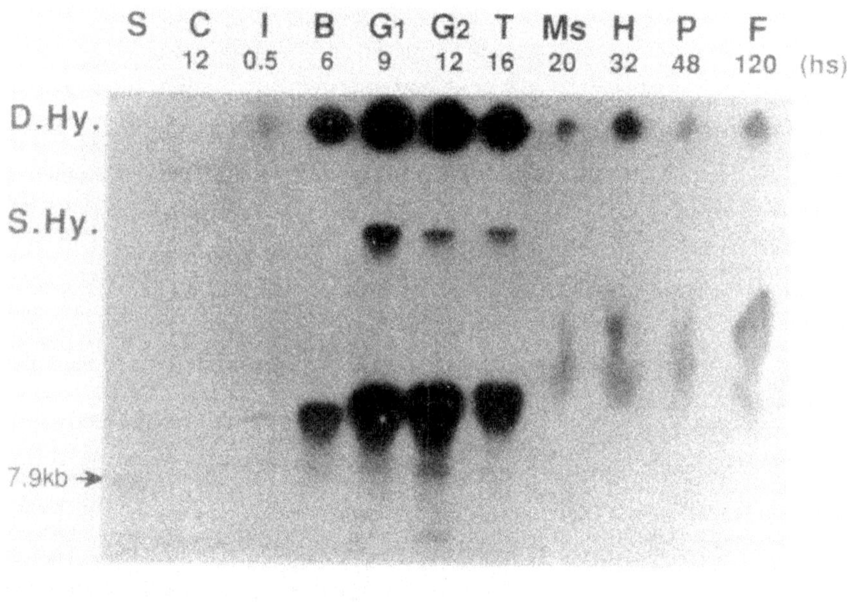

Fig. 4. Behaviour of the pMhGH gene (7.9 kb, see Fig. 2) during the host loach embryogenesis. DNA samples from each developmental stage were hybridized with the ^{32}P-labelled hGH gene probe. D. Hy., dot blot hybridization of DNA equivalent to 2 embryos; S. Hy., Southern blot hybridization of DNA equivalent to 4 embryos; S, 7.9 kb linear fragment of pMhGH gene, as injected; I, one cell stage (30 minutes after microinjection); B, blastula stage (6-hour embryos); G_1 and G_2, early and late gastrula stage (6-hour and 9-hour embryos, respectively); T, tail-bar stage (16-hour embryos); Ms, muscular reaction stage (20-hour embryos); H, heart beating stage (32-hour embryos); P, pigment forming stage (48-hour embryos), F, feeding stage (120 hour after fertilization) and C, control sample from non-injected G_2 stage.

of them were digested with either the restriction enzyme Sac I or Kpn 1. Southern blotting and hybridization with a ^{32}P-labeled probe of the hGH gene indicated that the portion of non-enzyme treated DNA sample showed bands which co-migrated with the high molecular weight cellular DNA at the top of the gel, while the Sac I digested DNA sample located at the 5.7 kb position, and the Kpn I digested sample showed bands intermediate between these positions. Since there was no Kph I restriction site within the MhGI/Sac I fragment,this result was interpreted as indicating that transgenes had randomly inserted into chromosomal DNA, maintaining the Sac I restriction site at both ends (Zhu et al., 1989). We assume that most of the transgenes were integrated as concatamers or polymers at late stages of embryogenesis, even though there may have been a few single copy insertions at earlier stages. During development, the transgenes should spread through the different tissues and organs of the transgenic fish. If the transgenes integrate into the DNA of gonadal cells, they would be passed on to next generation by germ line transmission.

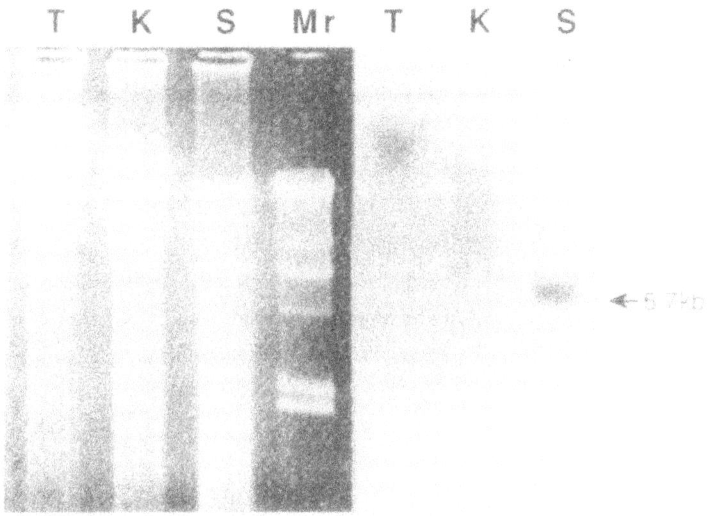

Fig. 5. Foreign gene integration in red crucian carp. DNA extracted from muscle of a red crucian carp with the MhGI transgene (5.7 kb/ Sac 1 fragment, see Fig. 2). Left: 0.7% agarose gel electrophoresis, 10 μg DNA for each lane. Right: Southern blot hybridization with the 32P-labeled hGH gene probe. T, DNA no restriction enzyme *treatment;* K, DNA with Kpn 1 digestion; S, DNA with Sac 1 digestion; Mr, molecular weight marker of l DNA/ Hind 111).

TRANSGENE EXPRESSION IN TRANSGENIC FISH

Gene expression was studied along with the identification of foreign gene integration. Samples of total nucleic acid were extracted from progressive developmental stages of groups of loach embryos which were microinjected with the linear pMhGH fusion gene. Agarose gel electrophoresis of the total nucleic acids revealed existence of 28S, 18S, and 5S RNAs at all developmental stages. In addition, an increasingly obvious smear of RNAs among 28S, 18S, and 5S RNAs also appeared. Blot hybridization with the 32P-labeled hGH gene probe indicated abundantly high gene transcripts at the muscular reaction stage (two bands) and a reduced amount at the feeding stage (one band) (Fig. 6).

Recently using Northern blot analysis, we detected multimer transcripts of the pMhGH gene during the blastula to muscular reaction stages of red crucian carp embryonic development. No hGH gene transcription was observed in the group in which the circular pMhGH gene had been transferred (data not shown). Prior to the muscular reaction stage, most copies of the foreign gene were concatamers present as extrachromosomal DNA.

The factors responsible for activating endogenous genes during cell differentiation

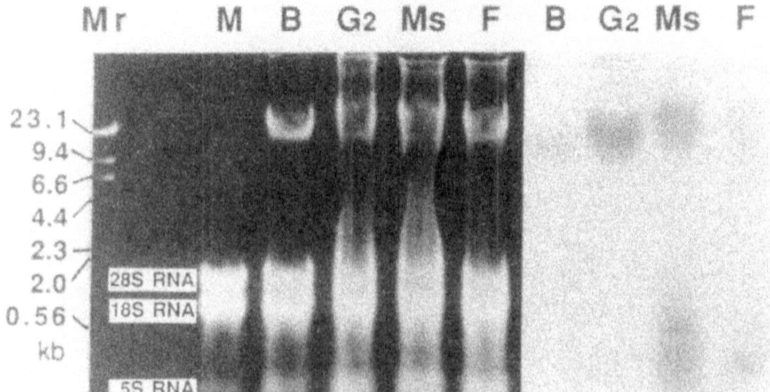

Fig. 6. Total nucleic acid Southern blot. The total nucleic acid was extracted from different stages of loach embryos containing the pMhGH transgene. Left: 0.7% agarose gel electrophoresis; Right: the gel blot hybridized with the 32P-labeled hGH gene probe. (Mr, DNA marker of λ/ Hing 111; M, multicellular stage; B, blastula stage; G2, late gastrula stage; Ms, muscular reaction stage; F, feeding stage).

may also activate the expression of non-integrated copies of the foreign gene. Interestingly, the MhGH gene copy number, transcript amount, and the expressed hGH content, at the heartbeating stage in abnormal host embryos were 2-3 times higher than in morphologically normal ones (Zou et al., 1990, in press) (Table 3). Extra copies of foreign gene replication and/or excessive expression (both the transcripts and hGH) interrupted or damaged the normal process of embryogenesis. After the feeding stage, on the other hand the foreign genes had been integrated into the host fry. Detectable transcripts at this stage originated from copies of the transgene integrated into the host chromosomes.

One advantage of choosing the human growth hormone gene as a model to study the generation of transgenic fish is that there is no cross-reactivity between the fish growth hormone. An ^{125}I-labeled monoclonal antibody against human growth hormone, routinely used in medical diagnosis, was obtained from the Beijing Institute of Neurosurgery. Twenty transgenic common carp were sacrificed at 35 days to obtain muscle samples, and 8 transgenic adult common carp were bled from the abdominal vein at 913 days. The results of the radioimmunoprecipitation assays showed that 10 out of 20 larval fish and 3 out of 8 adults expressed human growth hormone (Table 4). Although it was impossible to obtain serum samples from larval fish, it was clear that the expression level of the MhGH gene in young fish was much higher than that of

Table 3. The integration and expression of the pMhGH gene in heart beating stage embryos of crucian carp.

Morphological appearance	Copy no. of pMhGH/ embryo	Transcripts of pMhGH/ 50 embryos*	Content of hGH ng/embryo
Normal	1.0×10^7	1.0	2.0×10^{-2}
Abnormal	7.2×10^7	2.4	5.8×10^{-2}
Control	0	0	1.0×10^{-2}

* calculated by laser densitometer (LKB) of Northern blot autoradiograph

Table 4. Human growth hormone detected in transgenic common carp by radioimmunoprecipitation assay.

Age (days)	No. of fish tested	No. of fish with hGH	Range of hGH levels
35	20	10	26.1-50 ng/ fish (0.2 g)
913	8	3	1.1-4.0 ng/ ml serum

mature individuals. The developmental dynamics of hGH gene expression in fish is similar to that in humans, even though the transgenes were driven by a mouse MT-1 gene promoter and control region.

BIOLOGICAL FUNCTION OF HUMAN GROWTH HORMONE IN FISHES

A number of fish species have been experimentally treated with pituitary extracts or purified growth hormones from several sources including bovine (Higgs et al., 1975; Markent et al., 1977), porcine (Kormourdjian et al., 1978) and piscine (Adelman, 1977). All experiments resulted in increased growth rates. Biosynthetic recombinant chicken, bovine (Gill et al., 1985) and rainbow trout (Agellon et al., 1988) growth hormone have also been shown to enhance fish growth rates. Therefore it is reasonable to expect that human growth hormone expressed by a transgene in fish should function in the same way.

We have observed enhanced growth rates in MhGH gene- transferred loach and crucian carp (Zhu et al., 1986, 1989) (Fig. 7). During the first five feeding days the transgenic loach appeared smaller than controls, an observation that could be attributable to mechanical damage of the embryo at the time of microinjection. Two weeks after feeding the control fish were of uniform size, while the transgenic siblings showed considerable variation, with approximately 10% of them being about 1.5 times larger than the controls. A few of the transgenic individuals showed dramatic growth enhancement. For example at 43 days of age the average body weight of 12 control individuals was 35.1 mg (range 25 to 46 mg) while in the transgenic group there were two fry weighing 160 mg and 120 mg which is 4.6 and 3.4 times larger, respectively, then the average weight of the controls. In another experiment nine control juvenile loach weighed 0.77 g (range 0.7 to 0.9 g) on average at 135 days, whereas three transgenics weighed 2.78 g, 2.2 g and 2.1 g, and 3.8, 3.2 and 3.1 times as large, respectively, as the controls. These results strongly suggest that the growth enhancement observed in the loach was a function of the MhGH transgene product.

In the following two years, growth enhancement by the MhGH transgene was confirmed in silver crucian carp and common carp (Zhu et al., 1989). At 208 days of age transgenic silver crucian carp were 78% larger (weight) than controls. The largest transgenic carp (380 g) was 2.1 times heavier than the largest control. Similar results were observed for the common carp where at 153 days of age, transgenics were on average 9.4 % heavier than controls and the largest transgenic (330 g) was 1.4 times that of the largest control (240 g).

Parallel experiments with periodical intramuscular injections of biosynthetic human growth hormone (offered by Professor L. Guo, Shanghai Institute of Cell

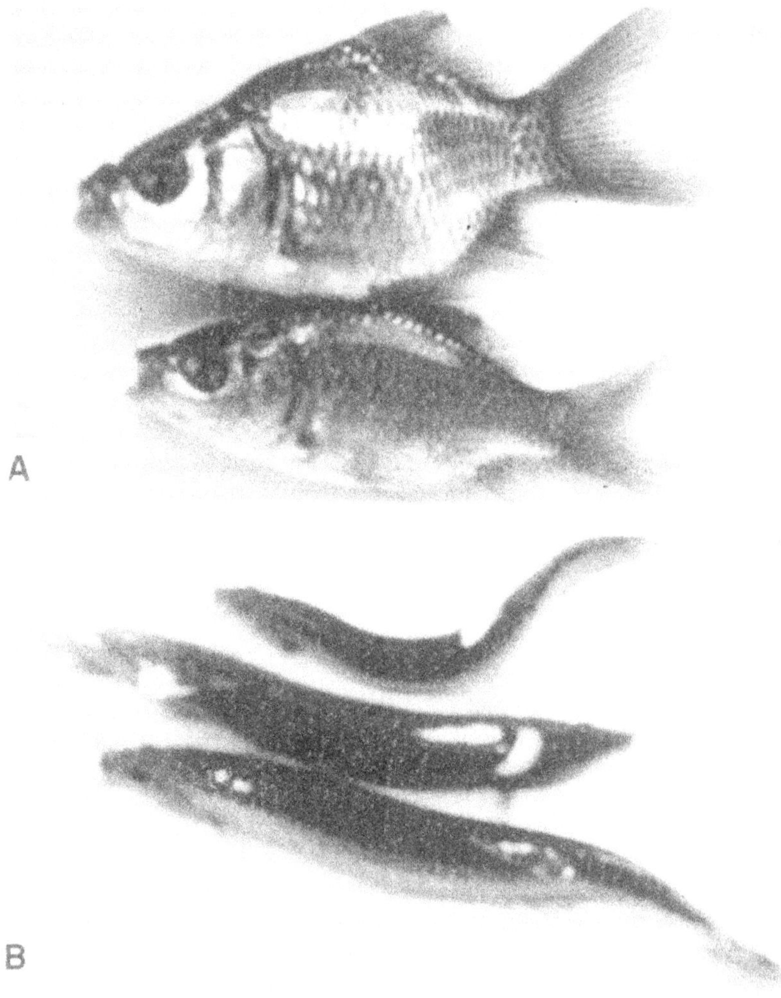

Fig. 7. The growth enhancement and morphological changes of transgenic crucian carp and loach. A, one transgenic crucian carp (upper, with *pMhGH* gene) and its non-transgenic sibling at 180 days; B, two transgenic loach (lower, with pMhGH gene) and their non-transgenic sibling at 135 days.

Biology, Chinese Academy of Sciences) into silver crucian carp were carried out by Xu et al., (1990). Two groups of 12 juveniles (60 days old) received hGH treatment at 2 µg/ g body weight/ week (group 2) and 10 µg/ g body weight/ week (group 3) for 6 weeks. One month after treatment, the average body weight of group 3 was 32.8% larger than the group 1 (the control group) (30.5 g/ 22.9 g) while the MhGH gene-transferred group was 17.7% larger than the group 1 (27.0 gl/22.9 g). There was no significant difference in body weight between group 2 and the controls (21.1 g/ 22.9 g) (Fig. 9). The dosage of hGH at 2 µg/ g body weight/ week may be too small to stimulate growth enhancement of silver crucian carp.

The mechanism of growth enhancement in the transgenic fish is thought to be based on the function of human growth hormone on skeletal enlargement and stimulation of protein syntheses. Some of the transgenic crucian carp were so increased in back muscle thickness and broadened in body width that they had a dramatically different morphology. Body width to body length is a common taxonomic characteristic. This figure in normal crucian carp is 35%-47% (Wu et al., 1977), in some of the transgenic crucian carp it increased to 51%-53.3% (Fig. 7: A). Some of the transgenic loach changed morphologically from long and thin (eel-like) to shorter, thick and brawly (snakehead fish-like) (Fig. 7: B).

TRANSGENE INHERITANCE BY GERM LINE TRANSMISSION

All transgenic crucian carp and common carp with hGH genes so far examined in our laboratory became sexually mature in one to two years. In 1986, six transgenic females and one transgenic male matured and reproduced naturally in an isolated cement pool. DNA samples from 30 individual offspring were probed with a ^{32}P-labeled hGH gene by dot blot hybridization. All offspring were positive for the transgene but the signals varied in intensity. Radioimmunoprecipitation assays indicated that one quarter of the fry synthesized detectable amounts of human growth hormone (Maclean et al., 1987). Since the offspring were generated from 6 transgenic maternal parents it was difficult to determine the foreign gene inheritance pattern in terms of copy numbers and loci of integration.

In another experiment, a transgenic female common carp was induced to spawn and the eggs were artificially fertilized with milt of a non-transgenic common carp. Nine out of 18 offspring (50 %) carried the hGH gene in various copy numbers. Surprisingly, one offspring harbored ten times as many copies of the hGH gene than its maternal parent. This is the first case we have found where a transgene appeared to be amplified in an offspring by germ line transmission. The mechanism accounting for this is unknown. At 7 months three "giant" juveniles emerged from a few hundred littermates in a small, well isolated fish pond. One " giant" was 7.7 times larger in body weight than the average weight of its siblings (Fig. 10).

DISCUSSION

The Fish Embryo as a Model System for Micromanipulation

Since T. C. Tung introduced the technique of nuclear transplantation to fish system, intensive studies with the technique of micromanipulation on the developmental

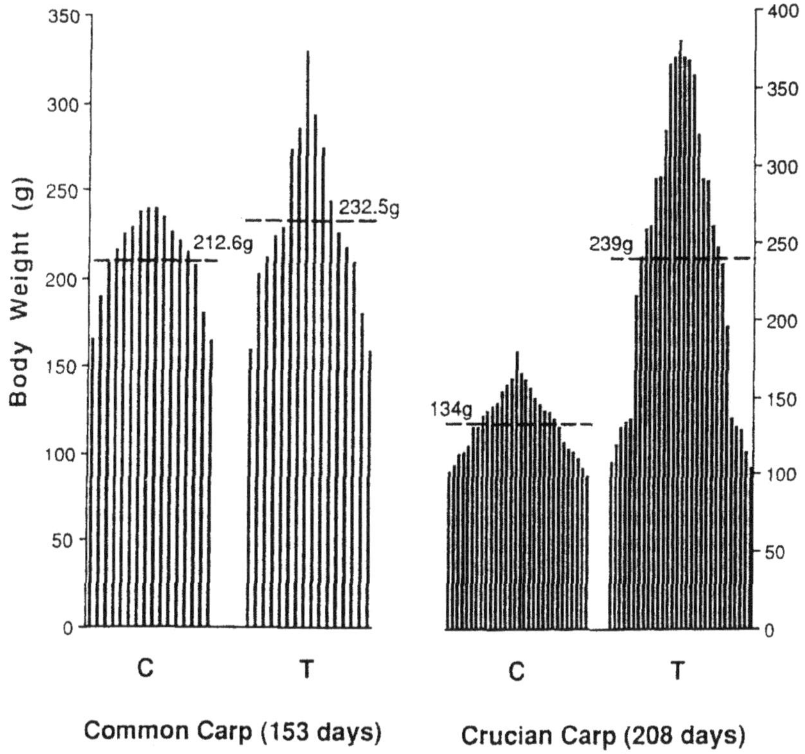

Fig. 8. The growth enhancement of hGH gene in transgenic common carp and crucian carp. Each bar represents an individual body weight. The dash line represents the average body weight of the group. C, control group, non-transgenic; T, experimental group with the pMhGH transgene.

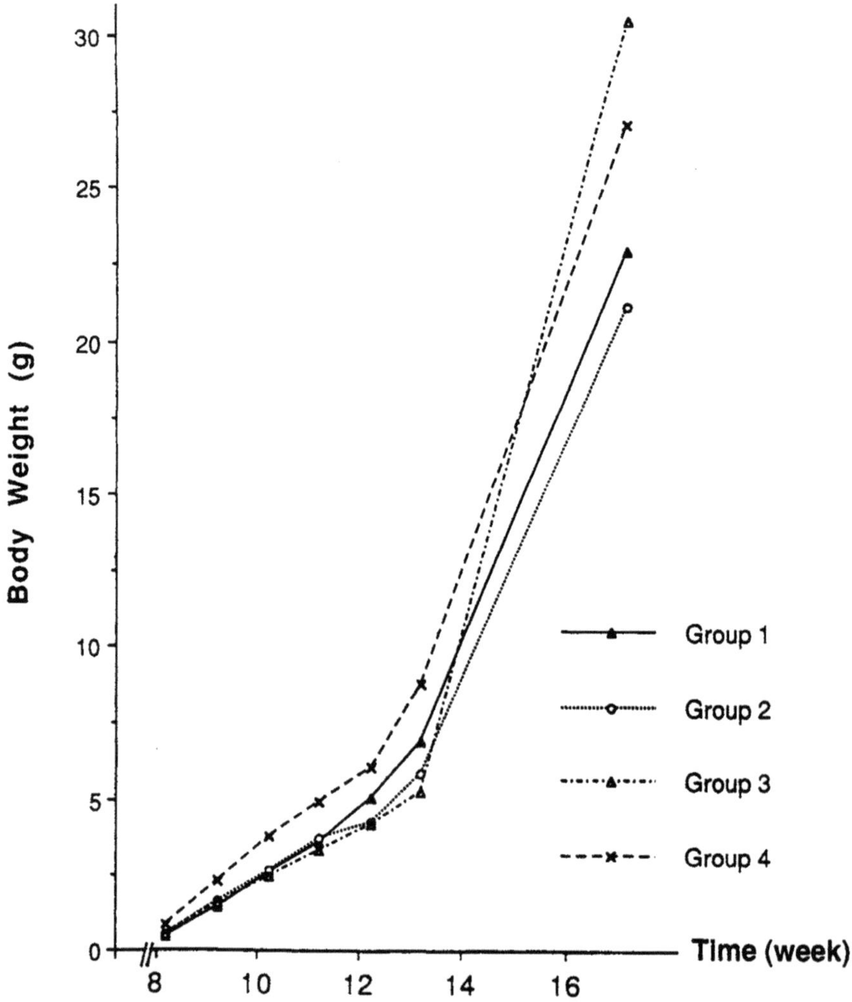

Fig. 9. The influence of hGH administration and hGH gene transfer on the growth rates of silver crucian carp. The comparison of the growth rates was made among four groups of 12 silver crucian carp which were siblings from the same parental fish. Group 1 was a control group without hGH administration and gene transfer. Group 2 and group 3 were subjected to intramuscular injection of hGH. The injection started at the age of 60 days and was carried out once a week for 6 weeks. The dosage of hGH in group 2 was 2μg/ g body weight/ week and in group 3 10 μg/ g body weight/ week. Group 4 was developed from the MhGH transgene containing eggs.

Fig. 10. An extreme example of segregation in transgenic carp offsprings. One offspring was 7.7 times larger in body weight than its siblings at 210 days of age.

biology of fish has been carried out. Fish eggs possess all of the experimental advantages attributable to amphibian eggs, such as inducible spawning and external fertilization. Eggs are numerous and in many species they are quite large. Fertilized eggs are easily maintained in water and the embryonic development of most species is rapid (Maclean et al., 1987, Powers, 1989). In some species, the egg chorion is easily removed by mechanical means (Tung et al., 1963), or enzymatic digestion (Zhu, 1982), (Hallerman et al., 1989) without interruption of embryogenesis. The embryo plate is transparent or translucent and the blastoderm is fairly distinguishable. In addition, the totipotency of a fish cell (both embryonic and somatic cells) is much higher than that observed for other vertebrate species. Nuclear transplantation can be successfully carried out between different species and even between different genera (Research group of Cytogenetics, 1980, Yan et al., 1984). The nuclei from over 300 days of subcultured blastula cells or from short-term cultured kidney cells, can promote embryogenesis and development in enucleated eggs of crucian carp (Evans and Kaufman, 1981). The combination of embryo-manipulation and cultured cell-manipulation in fish will certainly open a new window to developmental biologists, geneticists, and molecular biologists.

Production of a Pure Line of Transgenic Fish.

A pure line of transgenic fish means a cloned strain of fish in which all individuals (male and female) have identical copy numbers and integration sites of the transgene in their genome. In the mammalian system, an embryonic stem cell (or ES cell) system has been developed (Evans and Kaufman, 1981), which may eventually facilitate the creation of pure lines of transgenic mammals. The favorable features offered by fish eggs, indicated above suggest that the first successful creation of a pure transgenic line will be in fish rather than in any other vertebrate.

The strategy for producing a pure line of transgenic fish in our laboratory includes three steps. The first step is the generation of a totipotent transgenic cell in which all copies of the foreign gene(s) are integrated into the genome. One way to achieve this is to introduce cloned gene(s), with a reporter gene if possible, into the nucleus of cultured cells. After a few rounds of subculture, some cells will have integrated the foreign gene into their genome. The transgenic cells could then be selected out from among the non-transgenic ones. An alternative way is to transfer the gene directly into fertilized eggs and to isolate developing blastula cells for subculture, or/and subsequent selection for those cells containing the transgene. Whatever cultured cell or blastula cell is employed, the aim of this step is to ensure that all copies of foreign gene are stably integrated into the host cell chromosomes. The second step to generate homogeneous transgenic fish is by nuclear transplantation, i.e. replacement of the egg pronucleus by a totipotent transgenic cell nucleus. The procedure of serial nuclear transplantation is recommended to facilitate the adaptation of the totipotent transgenic cell nucleus to the cytoplasm of the enucleated egg. This also offers another chance to make additional stable transgenic cells. The third step is to mate homogeneous transgenic individuals to produce a pure line of transgenics. Multiple measures will be employed for breeding manipulation (Romashov and Belyaeva, 1964; Stanley, 1976; Liu et al., 1987) such like gynogenesis, androgenesis, sex reversal, classical selection and so on.

Site-Specific Integration

Although it is essential to have the foreign gene stably integrated into all of the host organisms cells, the specific site of its chromosomal integration may be important for proper expression of the desired phenotype. In MhGH gene-transferred loach a high number of abnormally developing embryos and high mortality rates at the heart beating stage were associated with an excess of the foreign hGH gene and/or its product (Table 4) (Zhu et al., 1990). In crucian carp, the growth rates of some of the individuals transgenic for MhGH were slower than those of controls. These deleterious phenomena could be associated with the site at which the foreign DNA integrated into the host chromosome(s).

There are three categories of integration sites (Zhu et al., 1986): (1) functional integration - foreign gene integration where it is suitable for expression, for example between the flanking regions of two functional genes; (2) silent integration - the foreign gene is integrated in a chromosomal region where it cannot be expressed, for example heterochromatin; (3) toxic integration - the foreign gene interrupts or breaks sequences critical to normal cellular function, such as those of so called "house-keeping' genes.

In our own studies, functional integration of the GH gene should result in the generation of fast growing fish. However over expression of GH may be damaging to the development and physiology of the fish. The abnormal development and slow growth observed in some of our transgenics could have been the result of a toxic integration and/or a functional integration with over expression of the GH.

Site-specific integration has not been attempted with fish to date. One approach currently being pursued is to ligate the foreign gene to specific sequences isolated from the host genome; for example flanking sequences from functional genes, or moderately repetitive sequences. Recently we have isolated and cloned two moderately repetitive sequences from the common carp genomic DNA. We hope to use these sequences to target foreign gene integration into functional regions of the host fish's chromosomes.

Suitable Expression and Growth Enhancement

The results of our studies indicate that the MhGH gene was expressed and accelerated the growth rates of individual transgenic loach, crucian carp and common carp. The proportion of transgenic individuals expressing the transgene varied with the fish species and with the experimental batch of fish. In addition the amount of expression varied between individuals. Furthermore the functional effects of the human growth hormone gene on the transgenic fish differed between species. In general, the smaller the full normal size of the fish, the more dramatic the growth effects of the transgene. Growth enhancement was also much greater during the fry to juvenile stages than it was during the adult stage. Therefore the growth hormone gene transfer could be most effective for increasing the full mature size of normally small but valuable species, or for shortening the duration it takes to produce market-size fish.

In order to reach the final goal of breeding a fast growing transgenic fish strain two questions need to be considered. (1) Why do some transgenics grow faster than others? (2) Is the MhGH gene really desirable for fish breeding? In our own research we believe that over expression of the hGH gene could have been responsible for the abnormal embryo development and mortalities observed. This implies that only a certain amount of growth hormone expression is required to maximize the growth rates of the fish. The expression level of a transgene will be determined by its integration site, copy number and promoter-control region. The results of our studies on the common carp transgenic for the hGH gene indicate that the mouse MT-1 promoter and control sequences do drive the expression of the hGH gene (Table 4), but at lower levels than those observed in transgenic mammals (Zhu et al., 1989). Heavy metal ions (zinc and cadmium) did not enhance the expression of the transgene in the fish. However even if the gene was inducible, toxic agents such as heavy metals should not be used in fish culture ponds. Therefore it will be necessary to look for a promoter other than that of metallothionein.

Human growth hormone does promote fish growth rates, but is it as efficient as a fish growth hormone? If so, are fast growing fish which contain a certain amount of human growth hormone, acceptable to consumers?

In consideration of the possibility that human growth hormone may not be acceptable, we have cloned the growth hormone and ß-actin genes of the common carp and grass carp (Liu et al., 1989, Zhu et al., 1990, Liu et al., submitted), and synthesized a new construct, pCAgGH (Zhu et al., unpublished data). The construct pCAgGH consists of the common carp ß-actin gene promoter and control region driving the grass carp GH genomic gene inserted into the plasmid pUC1 18. The common carp ß-actin gene promoter is a strong promoter and grass carp is a fast growing farmed species, it is reasonable to believe that the pCAgGH gene will be a strong "generator" to promote fish growth rates. Examination of pCAgGH gene expression is under-way.

ACKNOWLEDGMENTS

I gratefully acknowledge my research assistants and students, G. Li, L. He, Y. Xie, K. Xiu, Y. Wei, J. Zou, and D. Liu, for their common effort throughout the whole project. I would like to thank Dr. D.H. Hamer for the gift of the plasmid pBPVMG and Professor L.H. Guo for the hGH. This work was financially supported by the Chinese Academy of Sciences, the National Development Programme (the 7th five year plan) Research Fund, and the National High Technology and Development Programme Research Fund in China.

REFERENCES

Adelman, I.R. (1977) Effects of bovine growth hormone on growth of carp (*Cyprinus carpio*) and the influences of temperature and photoperiod. J. Fish Res. Board Can. 34: 509-515 .

Agellon, L.B., Emery, C.J., Jones, J.M., Davies, S.L., Dingle, A.D. and Chen, T.T. (1988) Promotion of rapid growth of rainbow trout (*Salmo gairdneri*) by a recombinant fish growth hormone. Can. J. Fish. Aquat. Sci. 45: 146-151.

Bending, M.M. (1981) Persistence and expression of histone genes injected into Xenopus eggs in early development. Nature 292: 65-67.

Brem, G., Brenig, B., Horstgen-Schwark, G. and Winnacker, E.L. (1988). Gene Transfer in Tilapia (*Oreochromis niloticus*). Aquaculture 68: 209-219.

Briggs R. and King, T.J. (1952) Transplantation of living nuclei from blastula cells into enucleated frogs' eggs. Proc. Natl. Acad. Sci. USA 38: 455-463 .

Bye, V.J. and Lincoln, R.F. (1986) Commercial methods for the control of sexual maturation in rainbow trout (*Salmo gairdneri*). Aquaculture 57: 299-309.

Chen, H., Yi, Y., Chen, M. and Yang, X. (1986) Study on the developmental potentiality of culture cell nuclei of fish. Acta Hydrobiologica Sinica 10(1): 1-7.

Chourrout, D. and Quillet, E., (1982). Induced gynogenesis in the rainbow trout: sex and survival of progenies. Production of all triploid populations. Theor. Appl. Genet. 63: 201-205.

Chourrout, D., Guyomard, R. and Houdebine, L. (1986) High efficiency gene transfer in rainbow trout (*Salmo gairdneri*) by microinjection into egg cytoplasm. Aquaculture 51: 143-150.

Constantini, F. and Lacy, E. (1981) Introduction of a rabbit beta-globin gene in the mouse germ line. Nature 294: 92-94.

Donaldson, E.M. and Hunter, G.A. (1983) Induced final maturation, ovulation, and spermiation in cultured fish. In: Fish Physiology. Hoar, W.S., Randall, D.J., and Donaldson, E.M. (eds.) Academic Press, New York. 9B: 405-435.

Dunham, R.A., Eash, J., Askins, J. and Townes, T. M. (1987) Transfer of the metallothionein-human growth hormone fusion gene into channel catfish. Trans. Am. Fish. Soc. 116: 87-91.

Etkin, L.D., Pearman, B., Robeils, M. and Bektesh, S.L. (1984) Replication, integration and expression of exogenous DNA injected into fertilized eggs of *Xenopus laevis*. Differentiation 26: 194-202.

Evans, M.J. and Kaufman, M.H. (1981) Establishment in culture of pluripotential cells from mouse embryos. Nature 292: 154-156.

Fletcher, G.L., Shears, M.A., King, M.J., Davies, P.L. and Hew, C.L. (1988) Evidence for antifreeze protein gene transfer in Atlantic salmon (*Salmo salar*). Can, J. Fish. Aquat. Sci. 45: 352-357.

Ge, W. and Jiang Y. (1985) A preliminary study on the mode in the egg of the naturally gynogenetic crucian carp for inhibiting heterologous sperm from transforming into male pronucleus. Acta Hydrobiologica Sinica 9(3): 203-208.

Gill, J.A., Sumpter, J.P., Donaldson, E.M., Dye, H.M., Souza, L., Berg, T., Wypych, J. and Langley, K. (1985) Recombinant chicken and bovine growth hormones accelerate growth in aquacultured juvenile Pacific salmon *Oncorhynchus kisutch* Bio/Technology 3: 643-646.

Gordon, J.W. and Ruddle, F.H. (1981) Integration and stable germ-line transmission of genes injected into mouse pronuclei. Science 214: 1244-1246.

Grunstein, M. and Hogness, D.S. (1975) Colony hybridization: A method for the isolation of cloned DNAs that contain a specific gene. Proc. Natl. Acad. Sci. USA 72: 3961-3965.

Gurdon, J.B. (1962) Adult frogs derived from the nuclei of single somatic cells. Biology 4: 256-273.

Hallerman, E.M., Schneider, J.F., Gross, M., Faras, A.J., Hackett, P.B., Guise, K.S. and Kapuscinski, A.R. (1989) Enzymatic dechorionation of goldfish, walleye, and northern pike eggs. Trans. Am. Fish. Soc. 117: 456-460.

Higgs, D.A., Donaldson, E.M., Dye, H.M. and McBride, J.R. (1975) A preliminary investigation of the effect of bovine growth hormone on growth and muscle composition of coho salmon (*Oncorhynchus kisutch*). Gen. Comp. Endocrinol. 27: 240-253.

Komourdjian, M.P., Burton, M.P. and Idler, D.R. (1978) Growth of rainbow trout, *Salmo gairdneri*, after hypophysectomy and somatotropin therapy. Gen. Comp. Endocrinol. 34: 158-162.

Liu, H., Yi, Y. and Chen, H. (1987) Generation of androgenetic homozygous diploid loach (*Misgurnus anguillicaudatus*). Acta Hydrobiologica sinica 11: 241-246.

Liu, Z., Zhu, Z., Roberg, K., Faras, A.J., Guise, K.S., Kapuscinski, A.R. and Hackett, P.B. (1989) The ß-actin gene of grass carp (*Ctenopharyngodon idellus*). Nucleic Acids Research 17(14): 5850.

Liu, Z., Zhu, Z., Roberg, K., Faras, A.J., Guise, K.S., Kapuscinski, A.R. and Hackett, P.B. (1991) Isolation and characterization of the ß-actin gene of carp (*Cyprinus carpio*) (submitted to Gene).

Maclean, N., Penman, D. and Zhu, Z. (1987a) Introduction of novel genes into fish. Bio/Technology 5: 257-261.

Maclean, N., Woodall, C., and Crossley, F. (1987b) Injection of the mouse MT-1 gene into rainbow trout eggs and assay of trout fry for resistance to cadmium and zinc. Experientia, Suppl. 52: 471-475.

Maniatis, T., Fritsch, E.F. and Sambrook, J. (1982) Molecular cloning--a laboratory manual. Cold Spring Harbor Laboratory Press, N-Y, pp. 89-94.

Markert, J.R., Higgs, D.A., Dye, H.M. and MacQuarrie, D.W. (1977) Influence of bovine growth hormone on growth rate, appetite, and food conversion of yearling coho salmon (*Oncorhynchus kisutch*) fed two diets of different composition. Can. J. Zool. 55: 74-83.

McEvoy, T., Stack, M., Keane, B., Barry, T., Sreenan, J. and Gannon, F. (1988) The expression of a foreign gene in salmon embryos. Aquaculture 68: 27-37.

Ozato, K., Kondoh, H., Inohara, H., lwamatsu, T., Wakamatsu, Y. and Okado, T.S. (1986) Production of transgenic fish: introduction and expression of chicken crystallin gene in medaka embryos. Cell Differ. 19: 237-244.

Palmiter, R.D., Brinster, R.L., Hammer, R.E., Trumbauer, M.E., Rosenfeld, M.G., Bimberg, N.C. and Evans, R.M. (1982) Dramatic growth of mice that develop from eggs microinjected with metallothionein-growth hormone fusion gene. Nature 300: 611615.

Pavlakis, G.N. and Hamer, D.H. (1983) Regulation of a metallothionein-growth hormone hybrid gene in bovine papilloma virus. Proc. Natl. Acad. Sci. USA 80: 397-401.

Pickford, G.E. (1954) The response of hypophysectomized male killfish to purified fish growth hormone as compared with the response to purified beef growth hormone. Endocrinology 55: 274-287.

Powers, D.A. (1989) Fish as model system. Science 246: 352-358.

Purdon, E.C. (1984) Atypical modes of reproduction in fish. In: Oxford reviews of reproductive biology. Clarke, J.R. (ed.) Oxford University Press, Oxford. 6: 303-340.

Research group of cytogenetics, Institute of Zoology, Academia Sinica, Research group of Somatic cell genetics, Institute of Hydrobiology, Academia Sinica, and research group of nuclear transplantation, Chang Jiang Fisheries Research Institute, (1980) Nuclear transplantation in teleosts--hybrid fish from the nucleus of carp and cytoplasm of crucian. Scientia Sinica 23(4): 517-525.

Rokkones, E., Alestrom, P., Skjervold, H. and Gautvik, K.M. (1989) Microinjection and expression of a mouse metallothionein human growth hormone fusion gene in fertilized salmonid eggs. J. Comp. Physol.(B) 158: 751-758.

Romashov, D.D. and Belyaeva, V.N. (1964) Cytology of radiation gynogenesis and androgenesis in the loach (*Misgurnus fossilis* L.) Dokl. Akad. Nauk. S.S.S.R. 157(4): 964-967.

Rusconi, S. and Scliaffner, W. (1981) Transformation of frog embryos with a rabbit beta-globin gene. Proc. Natl. Acad. Sci., USA 78: 5051-5055.

Stanley, J.G. (1976) Female homogamety in grass carp (*Ctenopharyngodon idellus*) determined by gynogenesis. J. Fish. Res. Board Can. 33: 1372-1374.

Stuart, G.W., McMurray, J.V., and Westerfield, M. (1988) Replication, integration and stable germ-line transmission of foreign sequences injected into early zebrafish embryos. Development 103: 403-412.

Thorgaard, G.H. (1986) Ploidy manipulation and performance. Aquaculture 57: 57-64.

Tung, T. C., Wu, S. C., Tung, Y. Y. F., Yan. S. Y., Tu, M. and Lu, T. Y. (1963) Nuclear transplantation in fish. Science Bulletin, Academia Sinica 7: 60-61.

Wagner, E.F., Stewart, T.A. and Mintz, B. (1981a) The ß-globin gene and a functional viral thymidine kinase gene in developing mice. Proc. Natl. Acad. Sci., USA 78: 5016-5020.

Wagner, T.E., Hope, P.C., Jollick, J.D., Scoll, D.R., Hodinka, R.L. and Galt, J.B. (1981b) Microinjection of a rabbit beta-globin gene into zygotes and its subsequent expression in adult mice and their offspring. Proc. Natl. Acad. Sci., USA 78: 6376-6380.

Wu, S.W. and Chung, L. (1964) Progress and achievement in the artificial propagation of four farm's fishes in China. Contributions at the 1964 Peking symposium, GEN: 160: 203-218.

Wu, S. W. et al., (1977) Cyprinid fishes of China (Vol.2), Shanghai people's press, Shanghai, China, pp. 430-436.

Xie, Y., Liu, D., Zou, J., Li, G. and Zhu, Z. (1989) Novel gene transfer in the fertilized eggs of loach via electroporation. Acta Hydrobiologica Sinica 13(4): 387-389.

Xu, K., Wei, Y., Guo, L. and Zhu, Z. (1990) The comparison of growth enhancement of crucian carp by hGH gene transfer and hGH administration. Acta Hydrobiologica Sinica 14 (in press).

Yan, S., Lu, D., Du, M., Li, G., Lin, L., Jin, G., Wang, H., Yang, Y., Xia, D., Liu, A., Zhu, Z., Yi, Y and Chen, H. (1984) Nuclear transplantation in teleosts--hybrid fish from the nucleus of crucian and the cytoplasm of carp. Scientia Sinica(B) 27(10): 1029-1034.

Yoon, S.J., Hallerman, E.M., Gross, M.L., Liu, Z., Schneider, A.J., Faras, A.J., Hackett, P.B., Kapuscinski, A.R. and Guise, K.S. (1990) Transfer of the Gene for neomycin resistance into Goldfish, Carassius auratus. Aquaculture 85: 21-23.

Zhang, P., Hayat, M., Joyce, C., Gonzalez-Villasenor, L.I., Lin, C.M., Dunham, R.A., Chen, T.T. and Powers, D.A. (1990) Gene transfer, expression and inheritance of pRSV-rainbow trout-GH cDNA in common carp, *Cyprinus carpio* (Linnaeus). Mol. Repr. Devel. 25: 3-13.

Zhu, Z. (1982) Embryonic development of *Claria fuscus* (Teleotei, siluriformes). Acta Hydrobiologica Sinica 7(4): 445-451.

Zhu, Z., Li, G., He, L. and Chen, S. (1985) Novel gene transfer into the fertilized eggs of goldfish (*Carassius auratus* L. 1758). Z. Angew. Ichthyol. 1: 31-34.

Zhu, Z., Xu, K., Li, G., Xie, Y. and He, L. (1986) Biological effects of human growth hormone gene microinjected into the fertilized eggs of loach *Misgurnus anguillicaudatus* (cantor). KeXue TongBao (Science Bulletin, Academia Sinica) 31(4): 988-990.

Zhu, Z., Xu, K., Xie, Y., Li, G. and He, L. (1989) A model of transgenic fish. Scientia Sinica (B) 2: 147-155.

Zhu, Z., He, L., Xie, Y. and Li, G. (1990) Construction of genomic library of carp (*Cyprinus carpio*) and grass carp (*Ctenopharyngodon idellus*) and isolation of their growth hormone genes and ß-actin genes. Acta Hydrobiologica Sinica 14(2): 176-178.

Zou, J., Xie, Y., Liu, D. and Zhu, Z. (1990) Foreign gene expression during fish embryogenesis. Acta Hydrobiologica Sinica 14 (in press).

CHAPTER 7

Expression of Heterologous Genes in Transgenic Fish

Boaz Moav, Zhanjiang Liu, Neomi L. Moav, Mark L. Gross, Anne R. Kapuscinski, Anthony J. Faras, Kevin Guise and Perry B. Hackett

ABSTRACT

Several studies have assessed the expression of foreign genes introduced into fish by microinjection of fertilized eggs. In recognition of the importance of tissue specificity and developmental stage specificity of several transcriptional control elements, an analysis of tissue and temporal specificity of expression from several transcriptional enhancer/promoter constructs has been initiated. Using the bacterial chloramphenicol acetyl transferase, and the bovine and chinook salmon growth hormone genes as reporter genes, the specificities of transcriptional control elements from the Rous sarcoma virus and carp ß-actin gene were analyzed during development in several species of fish including northern pike, walleye, and zebrafish. The effects of using either supercoiled or linearized recombinant constructs were examined.

INTRODUCTION

The multitude of factors controlling expression of foreign genes in transgenic animals is not fully understood. While the regulatory regions containing promoter and enhancer sequences are of critical importance to the control of gene expression (Nussinov, 1990), there are many cases where a regulatory region driving a transgene in one fish fails to do so in another, and often there is a considerable variation in the level of expression among animals carrying the same transgene (Krumlauf et al., 1985; Zhu et al., 1985, 1986; Chourrout et al., 1986; Maclean et al., 1987; Dunham et al., 1987; Enikolopov et al., 1988; Brem et al., 1988; Alok and Khillian, 1989; Guyomard et al., 1989; Rokkones et al., 1989). These observations suggest that several effects are involved in the control of transgene expression including the site of transgene insertion, effects of neighboring sequences, the number of transgenes inserted into the genome and availability of tissue-specific transcriptional factors. The integration of a normally active foreign gene into a transcriptionally inactive region of the mouse genome may result in a position-effect that modulates the effectiveness of the regulatory sequences of the transgene (Krumlauf et al., 1985 and Swift et al., 1984). Fifty percent of transgenic mice carrying the α-fetoprotein gene did not have a detectable level of transgene expression and the levels of expression in the other half of the animals were highly variable, indicating that the site of integration, as well as the promoter region was influencing the gene (Krumlauf et al., 1985).

Another factor thought to be involved in the regulation of foreign gene expression is either its methylation state or that of nearby regions of the host genome. Retroviral sequences undergo extensive de novo methylation after introduction into mouse embryos, which appears to correlate directly with gene inactivity (Palmiter et al., 1982). In addition, an insertional event can induce a change in the methylation pattern of the flanking host sequences, which then may exert a positional inactivating effect on transgene expression (Jahner, 1985). Genes transmitted through the female germline are often hypermethylated relative to the same gene transmitted through the male germline. Usually under-methylated genes are better expressed (Swain et al., 1987) due to differential binding of tissue-specific transcriptional factors (Iguchi-Ariga, 1989).

The transgene integration event must involve a rearrangement of the surrounding host DNA with or without modification of the transgene itself. This may affect the levels of expression of both the foreign and host gene sequences (Kratochwill et al., 1989). It has been demonstrated in several animals that multiple copies of the transgene may integrate into a single site as a unit, generally in a tandem head-to-tail array (Palmiter et al., 1982). However, expression is not always directly proportional to gene dosage (Palmiter et al., 1982).

a) Enhancer/Promoter Function In Vivo

Enhancer sequences lying several hundred base pairs (bp) upstream of a transcriptional start site can positively influence transcription in vivo, even when positioned in an inverted orientation in front of many different genes in various cell types of assorted organisms (Grosschel and Birnstiel, 1980, Banerji et al., 1981, Benoist and Chambon, 1981, Nussinov, 1990). The presence of introns within the injected transgene can contribute to the regulation of transgene expression, since animals with constructs carrying introns generally exhibit higher transcriptional rates than animals carrying identical genes lacking introns (Brinster et al., 1988). Some enhancing sequences may operate in a tissue-specific or cell type-specific manner. Banerji et al.,

(1983) have described IgG Kappa light chain enhancers that boost transcription of the ß-globin gene in an orientation- and position-independent manner in three lymphoid cell lines but not in brain cells. Gillies et al. (1983) have demonstrated that this orientation- and position-independent enhancer functions in lymphoid cells but not in L cells.

The number of identified tissue-specific enhancer elements is growing rapidly (e.g., Jones et al., 1988). The detailed characterization of DNA sequences exhibiting enhancer activity has led to the concept that an "enhancer or silencer effect" is produced by the combination of various sequence motifs or modules each of which contributes to the overall activity of the enhancer (Nussinov, 1990; Serfling et al., 1985). The enhancer sequence motifs thus far identified have generally been found to be binding sites for one or more transacting nuclear factors (Jones et al., 1988). The enhancer effect depends not only upon the mixture of sequence motifs comprising a particular enhancer sequence (Ondek et al., 1988), but also upon the assortment of transacting factors present within a particular cell type. A number of enhancer motifs can be recognized by multiple protein factors (Jones et al., 1988).

Chromatin changes induced by enhancers may be related to changes in superhelical conformation recognizable by transcription factors that result in transcriptional enhancement (Gilmour et al., 1986; Udvardy et al., 1989). Supercoiling in DNA greatly affects the expression of certain genes, including prokaryotic cistrons (Pruss and Drlica, 1989), 5S ribosomal RNA genes (Ryoji and Worcel, 1984), and SV40 genes on its minichromosome (Lucknik et al., 1982). Enhancers may exert their effects by acting as binding sites for transcription factors that interact with promoter-binding factors (Nussinov, 1990; Ptashne, 1988). Synergism between certain promoters and enhancers has been viewed as support for matched protein-protein interactions (Nussinov, 1990; Garcia et al., 1986).

The creation of transgenic fish with particular phenotypes depends on both the choice of gene and the specific transcriptional regulatory units responsible for the specificities in amount, tissue and developmental stage of expression. General considerations of transcriptional control units are discussed below.

b) The Role of Enhancers-Promoters in Tissue-Specific Expression

Three types of enhancers have been utilized in transgenic studies: (1) enhancers from normal cellular genes that are active in most if not all tissues, (2) enhancers active only in specific tissues, and (3) viral enhancers active in many tissues, especially those normally infected by the virus. In the following discussion we consider enhancer--

promoter complexes which normally act together (Lewin, 1990). The first category of enhancer-promoter complexes from cellular genes, active in a wide assortment of tissues and cells includes the metallothionein-1 (MT-1) (Durnam and Palmiter, 1981) and the ß-actin enhancer-promoter complexes (Liu et al., 1990). The MT-1 promoter has been found to direct expression of chimeric genes in a wide variety of tissues. The endogenous gene is normally expressed at moderate levels in the liver, kidney, gut and pancreas and at low levels in the brain and spleen (Durnam and Palmiter, 1981). MT-1 directed expression is not limited to its tissue of origin, indicating that expression of a transgene from the MT-1 promoter is susceptible to other influences, such as the site of integration. The ß-actin promoter, as described later, also has a wide spectrum of tissue activity. The second category of transcriptional regulatory units includes those taken from genes expressed exclusively in particular cell types but which can direct the expression of chimaeric fusion genes in transgenic tissues (see reviews Scangos and Bieberich, 1987; Rosenfeld et al., 1988; Gordon, 1989).

The third category of enhancers includes viral sequences which can direct expression of chimaeric genes in tissues normally infected by the virus. For example, the long terminal repeat (LTR) portion of the avian Rous sarcoma virus fused to the bacterial chloramphenicol acetyl transferase (CAT) gene (Gorman et al., 1982) directs expression in tissues of mesodermal origin such as muscle, tendon and bone (Overbeek et al., 1986). The LTR of the mouse mammary tumor virus (MMTV), with enhancer-promoter elements responsive to glucocorticoid regulation, directs expression of the thymidine kinase gene of *Herpes simplex* virus to lactating mammary tissue. However, viral LTR sequences are not limited to tissue-specific expression because other MMTV fusion genes can be expressed in additional tissues (Choy et al., 1987).

Tissue specificity is mainly due to the positive activation, and/or the negative repression, of transcription by transacting protein factors. Repressor elements can be located adjacent to or overlap the positive elements. The concerted influence of each element leads to differential expression and specificity (Hammer et al., 1987 and Lichtsteiner et al., 1987). Developmental control may simply be due to the loss of positive transacting regulatory factors.

One interesting aspect of gene regulation during development is that, as a tissue matures, some genes become susceptible to control mechanisms used at earlier developmental stages. This might be related to the fixation of chromatin in an "active" state wherein the enhancer is dispensable (Atchison and Perry, 1987), and/or permanently demethylated (Kelly et al., 1988).

Although it may be important, the question of tissue specificity of enhancer-promoter complexes has not yet been investigated in transgenic fish to any appreciable depth. We discuss below our initial studies in tissue-specific and developmental-stage-specific expression of transgene constructs of various enhancer-promoter units juxtaposed to several marker and test genes.

ANALYSIS OF TRANSGENE EXPRESSION IN FISH

a) Early Studies of Transgenic Fish

Several attempts have been made in the last few years to microinject genes into fish zygotes and to obtain transgenic fish (Zhu et al., 1985, 1986; Chourrout et al., 1986; Maclean et al., 1987; Enikolopov et al., 1987; Dunham et al., 1987; Brem et al., 1988; Alok and Khillian, 1989; Guyomard et al., 1989; Hew, 1989; Ozato et al., 1989; Powers, 1989; Fletcher and Davies, 1991; Gross et al., in press). In some of these experiments integration of the transgene into the fish genome was demonstrated, with occasional transfer of the transgene to offspring (Table 1). However, in general insufficient information has been obtained regarding regulation of expression of the transgene in the host fish (see Table 1).

Zhu et al. (1985, 1986) were the first to obtain transgenic fish. However, no direct data was presented demonstrating expression of the growth hormone gene (GH) driven by the mouse MT-1 (mMT-1) promoter. From the relatively large size of the F_1 offspring of these fish, Powers (1989) suggested the fish carried and expressed the GH gene. Zhu and coworkers (Maclean et al., 1987) have also demonstrated an increased average body weight of the transgenic loach (*Misgurnus sp.*) microinjected with human growth hormone driven by the mMT-1 promoter (see also Zhu, 1992, Chapter 6, this monograph).

McEvoy et al., (1988) have shown elevated expression of the bacterial lacZ gene product, ß-galactosidase (ß-gal) in salmon embryos microinjected with a mMT-1 promoter/lacZ gene construct. The human GH (hGH) driven by the mMT-1 promoter has been expressed in transgenic rainbow trout (rt) and Atlantic salmon (as) during early development [up to 42 days past fertilization, equivalent to the gastrula stages (Rokkones et al., 1989)]. Transcription of both the supercoiled and the linear plasmid was demonstrated, with a higher level of expression from the supercoiled plasmid. Rokkones et al., (1989) also demonstrated translational products from the transgene; the

Table 1. Gene expression in transgenic fish.

Structural Gene	Enhance Promoter	Evidence for Fish Species	Expression	References
Vertebrate genes of cDNAs				
hGH cDNA	mMt-1	carp	-	Alok and Khillian, 1989
hGH cDNA	SV40	trout	-	Chourrout et al., 1989 Guyomard et al., 1989
hGH cDNA	mMT-1	trout	-	Maclean et al., 1987
hGH cDNA	mMT-1	loach	+	Maclean et al., 1987 Zhu et al.,1985, 1986, 1991
hGH cDNA	mMT-1	loach	+	Enikolopov et al., 1986
hGH cDNA	mMT-1	trout	+	Rokkones et al.,1989
		Atlantic salmon	+	
hGH cDNA	mMT-1	catfish	-	Dunham et al., 1987
hGH cDNA	mMT-1	tilapia		Brem et al., 1988
bGH	RSV LTR	walleye	-	(this paper)
		northern pike	+	Gross et al., (in press)
chicken δ crystalin	chicken δ crystalin	medaka	+	Ozata et al., 1986 Inoue et al., 1989
Fish genes				
rtGH cDNA	RSV LTR	carp	+	Zhang et al., 1990
csGH cDNA	fish ß-actin	walleye		
		northern pike	+	(this paper)
anti-freeze	flounder anti-freeze	Atlantic salmon	+	Fletcher et al., 1988, 1991
GHcDNA	ocean pout antifreeze	Atlantic salmon	+	Du et al., 1991
Bacterial genes				
ß-gal (E. coli)	mMT-1	salmon	+	McEvoy et al.,1988
neo (E. coli)	RSV LTR	goldfish	+	Yoon et al., 1990 Guise et al., (1991)
Hygromycin (E. coli)	SV40	zebrafish	-	Stuart et al., 1988
CAT	fish ß-actin	zebrafish	+	Liu et al., 1990
		goldfish	+	Liu et al., 1990
		salmon	+	Liu et al., 1990
		walleye	+	Liu et al., 1990
		northern pike	+	Liu et al., 1990
CAT	RSV LTR	walleye	+	(this paper)
		northern pike	+	(this paper)
		goldfish	+	Hallerman et al., 1990
CAT	RSV/SV40	zebrafish	+	Stuart et al., 1990
		medaka	+	Chong & Vielkind, 1989

hGH was detected in the egg incubation water four days after microinjection. Early expression of transgenes may be due to transient expression from microinjected DNA rather than from mMT-1/hGH DNA integrated into the genome. Although reduced expression could result after integration because of position effects or methylation, expression during early development can be a good indication of the potential for expression of a particular construct in adult transgenic fish. For discussion, see Vielkind, 1992, Chapter 5, this monograph). Zhang et al. (1990) demonstrated the expression of a modified rainbow trout GH (rtGH) gene, in the red blood cells of transgenic carp. These carp had a significantly greater mean weight than control fish. Some of their progeny showed evidence for the presence of the introduced gene as well as an increased mean weight, suggesting that the rtGH cDNA was expressed in the F_1 generation (Chen et al., 1992, Chapter 9, this monograph). In summary, a multiplicity of transcriptional regulatory units from land vertebrates have been shown to be active in transgenic fish.

b) Initial Studies of Tissue Specificity of the RSV LTR in Transgenic Fish

The Rous sarcoma virus (RSV) LTR was used to drive bovine growth hormone cDNA (bGH) in transgenic walleye and northern pike. In this review we designate recombinate constructs by enhancer-promoter/gene, e.g., RSV-LTR/bGH. The RSV/bGH construct was co-microinjected with a marker construct RSV/CAT, in equal concentrations into the fish zygotes immediately after fertilization. The RSV/bGH construct promoted growth in some of these fish (Gross et al., in press). The RSV/CAT permitted an easy and rapid assay of expression during early development.

The results are shown in Tables 2 and 3. In northern pike embryos maintained at 11°C, expression of the RSV/CAT construct was relatively high around the time of hatching (about 10 days) and diminished thereafter. bGH was found in very early embryos (1 or 2 days of development) and occasionally in young fish. The RSV LTR was also active during early development in walleye embryos (Table 3). but we could not detect bGH in the walleye serum extracts taken between 13 and 36 days of development. Radioimmunoassays for bGH in serum of adult northern pike (10-12 months of age) obtained from the same population could only detect bGH in the blood of 49 individual fish out of 1,271 (3.8%) (Table 4).

These results indicated that the strong RSV LTR enhancer-promoter could drive the expression of heterologous mammalian and bacterial genes in several tissues of fish. Hallerman et al., (1990) reported the expression of RSV/CAT constructs in goldfish.

Hallerman et al., (1990) reported the expression of RSV/CAT constructs in goldfish. Fourteen of 20 mature fish tested were shown to have CAT activity, most frequently and at the highest levels in muscle. This tendency of enhanced expression in some tissues

Table 2. Expression of RSV promoters during early development of transgenic northern pike

Approximately 30 linearized RSV/CAT and RSV/BGH constructs (12.5 µg/ml each), were microinjected together into northern pike zygotes: embryos were divided into two equal samples for the measurement of CAT activity by (TLC) and bGH by specific RIA.
Embryos were homogenized with a Polytron homogenizer for 30 sec. in 0.25 M Tris (pH 7.4) at 4°C. The homogenates were then frozen in dry ice/ethanol for 5 min., and thawed at 37°C. The freeze-thaw procedure was repeated three times. The lysates were centrifuged in a microfuge for 5 min. The supernatants were used for CAT determinations (Gorman et al., 1982). After autoradiography, the spots were cut out and radioactivity was determined by scintillation counting and calculated as percent chloramphenicol converted into acetylchloramphenicol per hour per embryo. For RIA assays, embryos were homogenized in GH extraction buffer (50 mM Tris pH 8.5 0.1% Tween-20) extracted for 2 hr. at 4°C, then centrifuged for 90 min. at 15,000 g at 4°C. The supernatant was dialyzed overnight against 10 mM NaHCO3 and lyophylized. RIAs were performed by dissolving the powder in 0.1 ml of RIA buffer, 0.1 ml of antibody (titrated and diluted) and 0.1 ml of standard bGH (NIH), iodinated with ^{125}I (approximately 20,000 cpm). The mixture was incubated overnight at 4°C, 0.1 ml of Staph A was added and the mixture incubated for 2 hr. at 37°C. One ml of washing buffer was added, mixed and centrifuged for 30 min. at 4,000 g at 4°C. The pellet was counted with a gamma counter.

Days Post Injection	% Conversion Per Embryo	Extracts ng/Embryo	Days Post Injection	% Conversion Per Embryo	Extracts ng/Embryo
1	N.D	0.8	19	1.0	N.D
2	N.D	1.6	20	1.4	N.D.
3	N.D	N.D	21	8.8	----
4	3.0	N.D	28	2.2	----
5	5.8	N.D	33	N.D	----
6	N.D	N.D	40	N.D	(90/115)[a]
7	N.D	N.D	41	N.D	--(0/5)
8	9.0	N.D	47	N.D	--(0/4)
9	18.4	N.D	49	N.D	--(0/4)
10	15.3	N.D	50	N.D	--(0/6)
11	13.6	N.D	55	N.D	--(0/4)
12	17.5	N.D	57	N.D	--(0/4)
13	23.2	N.D			
14	45.0	-----			
15	11.2	N.D			
16	N.D	N.D			
17	N.D	N.D			

[a] Numbers of individual fish tested and found positive or negative, less than 0.1 ng bGH/embryo.
N.D not determined.

correlates with the expression of Rous sarcoma virus in avian and mammalian tissues (Overbeek et al., 1986).

The possibility that the RSV enhancer-promoter would drive the expression of transgenes in all tissues led Yoon et al., (1990) to transfer an RSV-bacterial neomycin phosphotransferance II (neo) gene into goldfish. The neo gene confers resistance to the neomycin analog G418 which has potential as a selectable marker in a wide range of eukaryotic organisms (Guise et al., 1992, Chapter 8, this monograph). RNA dot blot analysis indicated that half of the putative transgenic fish expressed the neo gene (Guise et al., 1991, Chapter 8, this monograph).

In summary the results obtained to date indicate that the RSV enhancer-promoter is expressed in a wide range of fish tissues.

ANALYSIS OF ß-ACTIN PROMOTER/ENHANCER IN TRANSGENIC FISH

a) The ß-actin Gene of Carp

The carp cytoplasmic ß-actin gene was chosen as a source for a potentially strong

Table 3. Expression of RSV/CAT in walleye during early development.

CAT activity was measured in embryo extracts by TLC and calculated as percent conversion of chloramphenicol per embryo per hour (see Table 2).

Days Post Injection	CAT % Conversion Per Embryo
1	--
2	--
3	1.5
4	3.5
5	N.D
6	N.D
7	7.0
8	9.8
9	3.5
10	6.3
11	5.5
12	2.3
13	0.3

-- less than 0.1% conversion per embryo per hour
N.D not determined

strong transcriptional promoter sequence to drive transgene expression in all cells of a transgenic fish. Liu et al. isolated and sequenced the ß-actin gene of the common carp (*Cyprinus carpio*) (Liu et al., 1990) and the grass carp (*Ctenopharyngodon idella*) (Liu et al., 1989). The two carp genes have identical sequences in regions implicated in the regulation of ß-actin gene expression. Consequently, we have examined the expression of the *Cyprinus carpio* ß-actin gene in tissue cultured mouse cells using the bacterial CAT reporter gene attached to various transcriptional regulatory elements from the carp gene. We constructed 30 different ß-actin/CAT constructs and, based on their expression in mouse cells, we (Liu et al., 1990) identified four transcriptional control regions in the carp ß-actin gene (Fig. 1): (1) A negative regulatory region 1-2 kb upstream of the gene. (2) A proximal promoter element containing the highly conserved CAAT and TATA boxes and the conserved CArG box found in all vertebrate ß-actin genes. (3) A negative element in the 5' region of the first intron. (4) A positive element near the 3' of the first intron. The positive regulatory element in the intron had position- and orientation-dependent effects on gene regulation. The various constructs containing the putative regulatory elements of ß-actin in different positions and orien-

Table 4. Expression of CAT and bGH genes in adult transgenic fish.

bGH was measured by RIA in serum as described in Table 2. Chinook salmon growth hormone (csGH) was measured by RIA in the serum [by the method of Wagner and Mckeown (1986)] of adult fish microinjected with the carp ß-actin gene enhancer-promoter (cßact, Rosenfeld et al., 1988,Liu et al., 1989, 1990, 1990, Moav et al., submitted) construct using an antibody against chum salmon GH.

			Number of fish with		
Fish Species	Construct	Number of Fish Tested	GH in Serum	CAT in Blood Cells	CAT in Fins
walleye	cßact/CAT or GH	125	3	--	25
northern Pike	RSV/CAT or GH	1,271	49	131/(6)[a]	(107)[b]/(3)[c]
	cßact/CAT or GH	171	N.D	12	56

a Number of CAT positives in blood among 49 fish with GH in the serum.
b Number of CAT positives in fins in the 131 fish with positive CAT in the blood.
c Number of CAT positives in fins among 49 fish with GH in the serum.
N-D not determined
– undetectable

tation were used to drive the CAT gene in three different fish cell lines: (1) EPC - epithelial carp cells, (2) RTH149 - rainbow trout hepatoma cells, and (3) RTG2 - rainbow trout fibroblasts. The proximal promoter element containing the TATA, CAAT and CArG boxes and the positive element containing a CArG box near the 3' end of the first intron were the major positive regulatory elements functional in tissue cultured fish cells (Moav et al., submitted). The "enhancer" element in the intron was only active in its original position and orientation, unlike a "classical" orientation- and position-independent enhancer.

Based on gel retardation assays, and "foot printing" experiments (Liu et al., in press) we mapped the location of one active element in the intron and found that both the proximal promoter and the intron-enhancer could compete for identical or similar nuclear

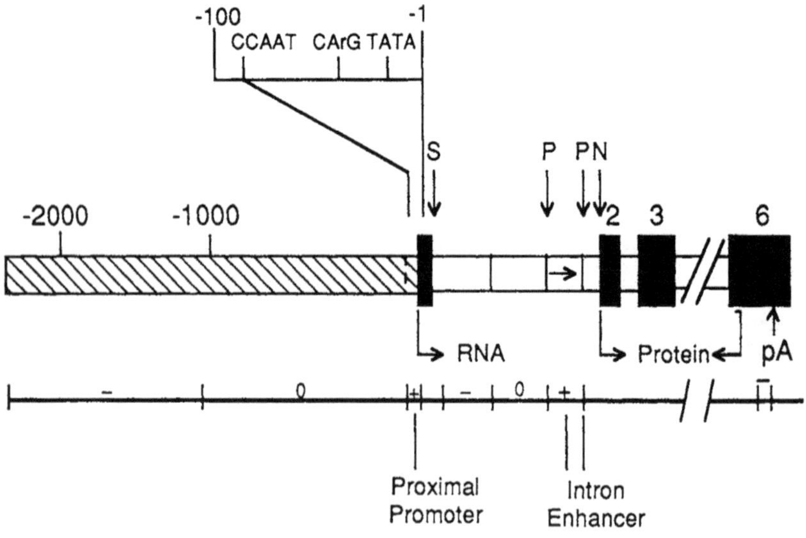

Fig. 1. Summary of the carp ß-actin gene regulatory regions. The map shows the ß-actin gene with the portion between exons 3 and 6 deleted. The first 100 bp of the proximal promoter element have been expanded. The site for initiation of transcription, the region that is translated, and the polyadenylation signal (pA) are designated a H, HindIII; P, PstI; R, EcoRV; S, SalI. The transcriptional regulatory loci identified in this paper are designated by + if positive, - if negative, and o if neutral. The negative regulatory region in exon-6 was mapped by Deponti-Zilli (1988) in the chicken ß-actin gene.

factors. These results suggested an interaction between the CArG elements in the proximal promoter and in the intron was required to achieve a high level of transcription.

b) Specificity of Expression of ß-actin/CAT Constructs

To confirm the usefulness of the carp ß-actin promoter (cßact) for driving transgene expression, we chose some of the constructs to test in transgenic fish (Tables 5 and 6). The cßact/CAT construct has 204 bp of the ß-actin proximal promoter region which contains the TATA, CAAT and CArG boxes (Fig. 1). This construct was microinjected into northern pike and walleye zygotes (Table 5). In both of these fish species the cßact/CAT construct had a unique temporal pattern of expression during early development. Compared to linearized DNA supercoiled plasmids had consistently higher levels of expression, with two peaks of activity in early development, and a longer persistence of activity. In the unintegrated state, supercoiled plasmids may either survive better in fish embryos or be replicated by fish DNA polymerase. A cßact/chinook salmon growth hormone cDNA (csGH) construct (Hew et al., 1989) produced detectable csGH in early development and in adult tissues by radioimmune assay (RIA) (Tables 4 and 6).

Expression of the cßact/CAT construct was detected in fin samples of 20% of adult walleye hatched from microinjected eggs. However, at the same time only 3 out of 125 showed evidence for salmon growth hormone in their serum (Table 4). In adult northern pike, about one-third (56/171) were positive for CAT following microinjection of cßact/CAT. In transgenic fish the ß-actin proximal promoter appears to be a more efficient promoter than the RSV LTR. The differences in activity may be due to differences in tissue specificity between the two enhancer-promoters.

The complete ß-actin regulatory complex (proximal promoter plus the intact first intron) produced higher levels of CAT activity than did the cßact/CAT construct both in cultured mouse and fish cells, transgenic northern pike, and transgenic walleye (Liu et al., 1990 and DePonti-Zilli et al., 1988). Similar conclusions were obtained following microinjection of the two CAT constructs into zebrafish zygotes (Liu et al., 1990, 1990b).

A construct that contains the proximal promoter of the carp ß-actin gene plus the enhancer element of the first intron in a reversed orientation was completely inactive in mouse and fish cells (Liu et al., 1990; Moav, submitted), but was surprisingly active during early development in walleye, northern pike and goldfish. In zebrafish, however, this construct was inactive (Liu et al., 1990). The patterns of expression of thirty ß-actin enhancer-promoter constructs in the various fish indicated that further analysis is needed.

Information gathered in the above experiments was used to develop "all-fish"

Table 5. Expression of linear and supercoiled cßact during early development of transgenic northern pike.

Microinjection of supercoiled (S.C) and linear plasmids were at a concentration of 12.5 μg/ml. Extracts in 0.25 mM Tris (pH 7.4) of embryos and fry were assayed for CAT activity and calculated as percent conversion of chloramphenicol per embryo per hour (see Table 2).

	CAT Activity Percent Conversion	
Days Post Injection	Per Embryo Linear	CAT Activity (S.C) Per Embryo
1	0.5	2.1
2	48.2	6.7
3	45.1	52.8
4	40.9	96.3
5	39.4	74.2
6	45.8	31.8
7	13.2	9.1
8	8.9	22.6
9	13.1	18.9
10	18.2	31.8
11	5.7	40.1
12	3.1	51.6
13	2.1	49.3
14	2.4	69.3
15	1.1	49.6
16	0.6	17.1
17	2.1	32.3
18	0.9	39.2
19	1.1	18.9
20	0.8	20.6
21	0.3	13.5
22	1.2	10.0
24	0.7	3.8
25	1.3	2.9
29	0.3	4.1 (1/6)[a]
36	0.6	10.4 (1/4)[a]

[a] The numbers represent the fraction of fry expressing CAT.

Table 6. Expression of supercoiled cßact/CAT and linear cßact/csGH during early development of transgenic walleye.

Either supercoiled or linear cßact/CAT plus linear cOact/csGH were microinjected. Linear cßact/CAT was injected together with linear cßact/csGH into the same zygotes. Embryonic extracts were divided into three equal samples and CAT and csGH activity was measured (see Tables 2 and 4).

	CAT Activity Percent Conversion		
Days Post Injection	Per Embryo Linear	Per Embryo S.C	csGH (RIA)
1	0.3	3.2	N.D
2	2.2	18.3	N.D
3	8.5	24.5	N.D
4	24.0	61.2	N.D
5	16.3	71.3	N.D
6	5.1	90.5	N.D
7	2.1	61.2	N.D
8	2.8	32.3	N.D
9	0.6	75.5	-
10	1.1	85.2	N.D
11	2.3	90.7	N.D
12	1.8	85.0	-
13	2.0	80.9	N.D
14	0.4	63.2	
16	1.8	45.1	+
18	--	27.6	+
20	--	6.3	+
22	0.3	3.3	-
23	1.6	2.9	-
25	2.2	10.6	-
27	--	1.9	-
28	1.2	6.1	-
29	1.6	2.1	-
31	7.5	1.7	+
34	0.3	3.1	-
35	0.3	1.1	-
37	0.8	2.6	-

N.D not determined
Negative (-) CAT results are less than O.1% conversion per embryo per hour. GH negative (-)are less than I ng per embryo above the background normal levels. GH positive (+) indicates a significantly higher level of csGH and walleye GH (which cross react with the anti-csGH A6) relative to GH levels in walleye which developed from control eggs.

expression vectors. One fish vector (FV-1) (Liu et al., 1990) contains the proximal promoter of ß-actin (Liu et al., 1990) the CAT gene as a reporter gene and the 3' region containing the polyadenylation signal from csGH gene (Hew et al., 1989). We did not use the polyadenylation site of the ß-actin since it apparently contains a conserved element that suppresses expression in muscle cells (Durnam and Palmiter, 1981 and DePonti-Zilli et al., 1988).

A second "all-fish" vector (FV-2) contains the whole enhancer-promoter complex of ß-actin (Liu et al., 1990), the CAT reporter gene and the 3' region of the csGH gene. A comparison, in tissue-cultured cell lines, of FV-1 and FV-2 to promoters from RSV, SV40 and the human heat shock 70 gene showed that the fish vectors were stronger in both mouse L cells and carp EPC cells. In transgenic zebrafish during early development, the FV-2 was superior to the other vectors tested and showed unique developmental patterns (Liu et al., 1990). In young transgenic goldfish these fish vectors also demonstrated a unique temporal pattern of expression (Fig. 2).

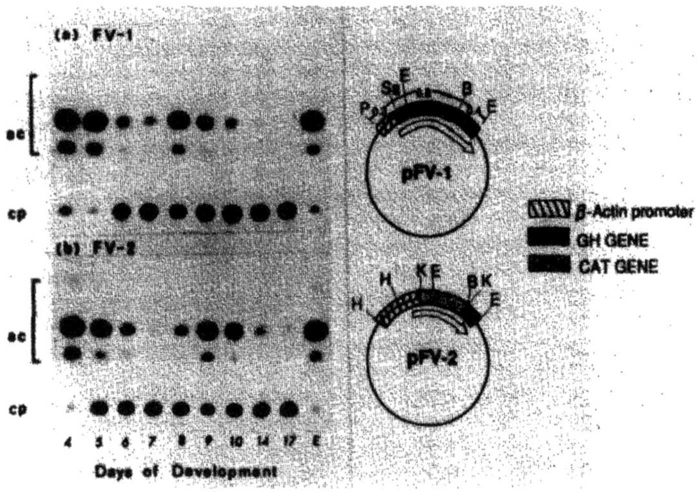

Fig. 2. Expression of fish vectors in goldfish. Goldfish zygotes were microinjected with pFV-1 and pFV-2. Extracts in 0.25 M Tris and CAT assays were performed as described in Table 2. Each lane represents the CAT activity in an amount of embryo extract equivalent to one embryo. cp,[^{14}C]-labelled chloramphenicol; ac, acetylated [^{14}C]-labelled chloramphenicol. Arrows in the schematics of the plasmid indicate the direction of transcription. B, Bgl; E, EcoRl; H, Hindlll; K. Kpnl: P [illegible] Sa, SalI. Sizes of units of pFV-1 are in kb.

In summary, studies of the specificity of promoter activity with respect to tissues and developmental stage have been initiated. The results have raised many questions regarding locational and temporal control of transgene regulation.

PERSPECTIVES

The ability to produce alterations in the phenotypes of transgenic fish in a functional and stable manner, offers the continued opportunity to study the regulation of gene function. A major remaining problem concerns the limits of independence of transgene expression. Clearly, we are just at the beginning of the study of gene expression in transgenic vertebrate animals in general and fish in particular. Future studies of gene expression in transgenic fish will require greater resolution of when, where and how transgenes are expressed in specific cells at given stages of development.

ACKNOWLEDGMENTS

This project was supported by the Minnesota Sea Grant Program supported by the NOAA (Office of Sea Grant, Department of Commerce under grant No. NOAA-86AA-D-SGI12-0 to P.H., paper JR265) and the Legislative Commission on Minnesota Resources to the Minnesota transgenic fish group (MTGF). Chinook salmon growth hormone cDNA was obtained from Choy Hew (Toronto) and salmon growth hormone in blood was assayed by Brian McKeown (Vancouver). We thank Luba Caldovic for helping with the screening of the adult transgenic fish, Kris Kirkeby and Mark Dalton for artwork, and Kae Ebling for typing the manuscript.

REFERENCES

Alok, D. and Khillian, J. S. (1989) Gene transfer in fish embryos for increased productivity. Theriogenology 31: 167.

Atchison, M.L. and Perry, R.P. (1987) The role of the K enhancer and its binding factor NF- KB in the developmental regulation of K gene transcription. Cell 48: 121-128.

Banerji, J., Rusconi, S. and Schaffner, W. (1981) Expression of a ß-globin gene is enhanced by remote SV40 DNA sequences. Cell 27: 299-308.

Banerji, J., Olson, L. and Schaffner, W. (1983) A lymphocyte-specific cellular enhancer is located downstream of the joining region in immunoglobulin heavy chain genes. Cell 33: 729-740.

Benoist, C. and Chambon, P. (1981) In vivo sequence requirements of the SV40 early promoter region. Nature 290: 304-310.

Brem, G., Brenig, G., Horstgen-Schwark, G. and Winnacker, E.L. (1988) Gene transfer in tilapia (*Oreochromis niloticus*). Aquaculture 68: 209-219.

Brinster, R.L., Allen, J.M., Behringer, R.R., Gelinas, R.E. and Palmiter, R.D. (1988) Introns increase transcriptional efficiency in transgenic mice. Proc. Natl. Acad. Sci. USA 85: 836-840.

Chen, T.T., Lin, C.-M., Dunham, R.A. and Powers, D.A. (1992) Integration, expression and inheritance of foreign fish growth hormone gene in trangenic fish. In: Transgenic Fish. C.L. Hew and G.L. Fletcher (eds.) World Scientific Publishing Co. Singapore, pp. 164-175.

Chong, S. S. C. and Vielkind, J. R. (1989) Expression and fate of CAT reporter gene microinjected into fertilized medaka (*Oryzias latipes*) eggs in the form of plasmid DNA, recombinant phage particles and its DNA. Theor. Appl. Genet. 78: 369-380.

Chourrout, D., Guyamard, R. and Houdebine, L.M. (1986) High efficiency gene transfer in rainbow trout. Aquaculture 51: 143-150.

Choy, Y., Henrard, D., Lee, I. and Ross, S.R. (1987) The mouse mammary tumor virus long terminal repeat directs expression in epithelial and lymphoid cells of different tissues in transgenic mice. J. Virol. 61: 3013-3019.

DePonti-Zilli, L., Seiler-Tuyns, A. and Paterson, B.M. (1988) A 40-base-pair sequence in the 3' end of the ß-actin gene regulates ß-actin mRNA transcription during myogenesis. Proc. Natl. Acad. Sci. USA 85: 1389-1393.

Du, S. J., Gong, Z., Fletcher, G. L., Shears, M. A. and Hew, C. L. (1992) Growth hormone gene transfer in Atlantic salmon: use of fish antifreeze/growth hormone chimeric gene construct. In: Transgenic fish. C. L. Hew and G. L. Fletcher (eds.) World Scientific Publishing Co. Singapore, pp. 176-189.

Dunham, R.A., Eash, J., Askins, J. and Towns, T.M. (1987) Transfer of the metallothionein-human growth hormone fusion gene into channel catfish. Trans. Am. Fish. Soc. 116: 87-91.

Durnam, D.M. and Palmiter, R.D. (1981) Transcriptional regulation of the mouse metallothionein-I gene by heavy metals. J. Biol. Chem. 256: 5712-5716.

Enikolopov, G.N., Benyumov, A. O., Barmintsev, V.A., Zelenina, I.A., Sleptsova, Yu. K., Doronin, V.A., Golichenkov, V.A., Grashchuk, M.A., Georgiev, G.P., Rubtsov, P.M., Skryabin, K.G. and Beav, A.A. (1988) Enhanced growth of transgenic fish containing human growth hormone gene. Doklady Akad. Nauk SSR 301: 724-727.

Fletcher, G.L., Shears, M.A., King, M.J., Davis, P.L. and Hew, C.L. (1988) Evidence for antifreeze protein gene transfer in Atlantic salmon (*Salmo salar*). Can. J. Fish. Aquat. Sci. 45: 352-357.

Fletcher, G. L. and Davies, P. L. (1991) Transgenic fish for aquaculture. In: Genetic Engineering, principles and methods. J. K. Setlow (ed.) Plenum Press, New York, pp. 331-370.

Fletcher, G. L., Davies, P. L. and Hew, C. L. (1992) Genetic engineering of freeze-resistant Atlantic salmon. In: Transgenic fish. C. L. Hew and G. L. Fletcher (eds.) World Scientific Publishing Co, Singapore, pp. 190-208.

Garcia, J.V., Bich-Thuy, L., Stafford, J. and Queen, C. (1986) Synergism between immunoglobulin enhancers and promoters. Nature 322: 383-385.

Gillies, S.D., Morrison, S.L., Oi, V.T. and Tonegawa, S. (1983) A tissue-specific transcription enhancer element is located in the major intron at a rearranged immunoglobulin heavy chain gene. Cell 33: 717-728.

Gilmour, D.S., Pflugfelder, G., Wang, J.C. and Lis, J.T. (1986) Topoisomerase I interacts with transcribed regions in Drosophila cells. Cell 44: 401-407.

Gordon, J. W. (1989) Transgenic animals. Int.Rev.Cytol. 115: 171-226.

Gorman, G.M., Merlino, G.T., Willingham, M.C., Pastan, I. and Howard, B.H. (1982) The Rous sarcoma virus long terminal repeat is a strong promoter when introduced into a variety of eukaryotic cells by DNA-mediated transfection. Proc. Natl. Acad. Sci. USA 79: 6777-6781.

Gross, M.L., Schneider, J.F., Moav, N., Alvarez, C., Myster, S., Liu, Z., Hew, C., Hallerman, E.M., Hackett, P.B., Guise, K.S., Faras, A.J., and Kapuscinski, A.R. (1991) Molecular analysis and growth evaluation of northern pike (*Esox lucius*) microinjected with growth hormone genes. Aquaculture (in press).

Grosschel, R. and Birnstiel, M.L. (1980) Spacer DNA sequences upstream of the T-A-TA-A-A-T-A sequence are essential for promotion of H2A histone gene. Proc. Natl. Acad. Sci. USA 77: 7102-7106.

Guise, K.S., Hackett, P.B. and Faras, A.J. (1992) Transfer of genes encoding neomycin resistance, chloramphenicol acetyl transferase, and growth hormone into goldfish and northern pike. In Transgenic fish. C. L. Hew and G. L. Fletcher (eds.) World Scientific Publishing Co, Singapore, pp. 142-163.

Guyomard, R., Chourrout, D., Leraux, C., Houdebine, L. M. and Pourrain, F. (1989) Integration and germ line transmission of foreign genes microinjected into fertilized trout eggs. Biochemie 71: 857-863.

Hallerman, E. M., Schneider, J. F., Gross, M., Liu, Z., Yoon, S. J., Hackett, P.B., Faras, A.J., Kapuscinski, A.R. and Guise, K.S. (1990) Tissue specific gene expression promoted by the RSV long terminal repeat element on transgenic goldfish. Anim. Biotech. 1: 79-93.

Hammer, R.E., Krumlauf, R., Camper, S.A., Brinster, R.L. and Tilghman, S.M. (1987) Diversity of alpha-fetoprotein gene expression in mice is generated by a combination of separate enhancer elements. Science 235: 55-58.

Hew, C.L. (1989) Transgenic fish: Present status and future directions. Fish Physiol. Biochem. 7: 409-413.

Hew, C.L., Trinh, K.Y., Du, S.J. and Song, S. (1989) Molecular cloning and expression of salmon pituitary hormones, Fish Physiol. Biochem. 7: 1-4.

Howlett, A.R., Cullen, B., Hertle, M. and Bissell, M.J. (1987) Tissue tropism and temporal expression of Rous sarcoma virus in embryonic avian limb in ovo. Oncogene Res. 1: 255-263.

Iguchi-Ariga, S.M. and Schaffner, W. (1989) CpG methylation of the cAMP-responsive enhancer/promoter sequence TGACGTCA abolishes specific factor binding as well as transcriptional activation. Genes Dev. 3: 612-619.

Inoue, K., Ozato, K., Kondoh, H., lwamatsu, T., Wakamatsu, Y., Fujita, T. and Okada, T.S. (1989) Stage-dependent expression of the chicken delta-crystallin gene in transgenic fish embryos. Cell Diff. Dev. 27: 57-68.

Jahner, D. (1985) Retrovirus-induced denovo methylation of flanking host sequences correlates with gene inactivity, Nature 315: 594-597.

Jones, N.C., Rigby, P.W.J. and Ziff, E.B. (1988) Transacting protein factors and the regulation of eukaryotic transcription: Lessons from studies on DNA tumor viruses. Genes Dev. 2: 267-281.

Kelly, D.E., Pollack, B.A., Atchison, M.L. and Perry, R.P. (1988) The coupling between enhancer activity and hypomethylation of K immunoglobulin genes is developmentally regulated. Mol. Cell Biol. 8: 930-937.

Kratochwil, K., von der Mark, K., Kollar, E.J., Jaenisch, R., Mooslehner, K., Schwartz, M., Haase, K., Gmachi, I. and Harbers, K. (1989) Retrovirus-induced insertional mutation in Movl3 mice affects collagen I expression in a tissue-specific manner. Cell 57: 807-816.

Krumlauf, R., Hammer, R.E., Tilghman, S.M. and Brinster, R.L. (1985) Developmental regulation of α-fetoprotein genes in transgenic mice. Mol. Cell Biol. 5: 1639-1648.

Lewin, B. (1990) Commitment and activation at PolII promoters: A tail of protein protein interactions. Cell 61: 1161-1164.

Lichtsteiner, S., Wuarin, J. and Schibler, U. (1987) The interplay of DNA-binding proteins on the promoter of the mouse albumin gene. Cell 51: 963-973.

Liu, Z., Zhu, Z., Roberg, K., Faras, A.J., Guise, K.S., Kapuscinski, A.R. and Hackett, P.B. (1989) Sequence of the grass carp ß-actin gene. Nucleic Acid Res. 17: 5850.

Liu, Z., Moav, B., Faras, A.J., Guise, K.S., Kapuscinski, A.R. and Hackett, P.B. (1990) Development of expression vectors for transgenic fish. Bio/Technology 8: 1268-1272.

Liu, Z., Zhu, Z., Roberg, K., Faras, A.J., Guise, K.S., Kapuscinski, Hackett, P.B. (1990) Isolation and characterization of the ß-actin gene of carp (*Cyprinus carpio*). DNA Sequence J. DNA Sequencing and Mapping 1: 125-136.

Liu, Z., Moav, B., Faras, A.J., Guise, K.S., Kapuscinski, A.R. and Hackett, P.B., (1990) Functional analysis of elements affecting expression of the ß-actin gene of carp. Mol. Cell Biol., 10: 3432-3440.

Liu, Z., Moav, B., Faras, A.J., Guise, K.S., Kapuscinski, A.R. and Hackett, P.B. (1991b) Importance of the CARG box in the regulation of ß-actin gene expression". Gene (in press).

Lucknik, A.N., Bakayev, V.V., Zbarsky, I.B. and Georgiev, G.P. (1982) Elastic torsional strain in DNA within a fraction of SV40 minichromosomes: Relation to transcriptionally active chromatin. EMBO J. 1: 1353-1358.

Maclean, N., Penman, D. and Zhu, Z. (1987) Introduction of novel genes into fish. Bio/Technology 5: 257-261.

McEvoy, T., Stack, M., Keane, B., Barry, T., Sreenan, J. and Gannon, F. (1988) The expression of a foreign gene in salmon embryos. Aquaculture 68: 27-37.

Moav, B., Liu, Z., Faras, A.J., Guise, K.S., Kapuscinski, A.R. and Hackett, P.B. (1990) Regulation of early expression of transgenes in developing fish. (submitted).

Nussinov, R. (1990) Signal sequences in eukaryotic upstream regions. CRC Crit. Rev. Biochem. Mol. Biol. 25: 185-224.

Ondek, B., Gloss, L. and Herr, W. (1988) The SV40 enhancer contains two distinct levels of organization. Nature 333: 40-45.

Overbeek, P.A., Lai, S.P., Van Quill, K.R. and Westphal, H. (1986) Tissue-specific expression in transgenic mice of a fused gene containing RSV terminal sequences. Science 231: 1574-1577.

Ozato, K., Kondoh, H., Inohara, H., Iwamatsul T., Wakamatsu, Y. and Okada, T.S. (1986) Production of a transgenic fish: Introduction and expression of chicken delta-crystallin gene in medaka embryos. Cell Diff. 19: 237-244.

Ozato, K., Inoue, K. and Wakamatsu, Y. (1989) Transgenic fish: Biological and technical problems. Zoological Sci. 6: 445-457.

Palmiter, R.D., Chen, H.Y. and Brinster, R.L. (1982) Differential regulation of metallothionein-thymidine kinase fusion genes in transgenic mice and their offspring. Cell 29: 701-710.

Powers, D.A. (1989) Fish as model systems. Science 246: 352 -358 .

Pruss, G.J. and Drlica, K. (1989) DNA supercoiling and prokaryotic transcription. Cell 56: 521-523.

Ptashne, M. (1988) How eukaryotic transcriptional activators work. Nature 335: 683-689.

Rokkones, E., Alestrom, P., Skjervold, H. and Gautvik, K.M. (1989) Microinjection and expression of a mouse metallothionein human growth hormone fusion gene in fertilized salmonid eggs. J. Comp. Physiol. B 158: 751-758.

Rosenfeld, M.G., Crenshaw, E.B., Lira, S.A., Swanson, L., Borrelli, E., Heyman, R. and Evans, R.M. (1988) Transgenic mice: Applications to the study of the nervous system. Ann. Rev. Neurosci. 11: 353-372.

Ryoji, M. and Worcel, A. (1984) Chromatin assembly studies in *Xenopus* oocytes: In vivo studies. Cell 37: 21-32.

Scangos, G. and Bieberich, C. (1987) Gene transfer into mice. Adv. in Genetics 24: 285-321.

Serfling, E., Jasin, M. and Schaffner, W. (1985) Enhancers and eukaryotic gene transcription. Trends Genet. 1: 224-230.

Stuart, G.W., McMurray, J.V. and Westerfield, M. (1988) Replication, integration, and stable germ-line transmission of foreign sequences into early zebrafish embryos. Development 103: 403-412.

Stuart, G. W., Vielkind, J. R., McMurray, J. V. and Westerfield, M. (1990) Germ-line transmission and variegated expression of a foreign gene in transgenic zebrafish. Development 109: 577-584.

Swain, J.L., Stewart, T.A. and Leder, P. (1987) Parental legacy determines methylation and expression of an autosomal transgene: A molecular mechanism for parental imprinting. Cell 50: 719-727.

Swift, G.H., Hammer, R.E., McDonald, R.J. and Brinster, R.L. (1984) Tissue-specific expression of the rat pancreatic elastase 1 gene in transgenic mice. Cell 38: 639-646.

Udvardy, A., Schedl, P., Sander, M. and Hsieh, T.S. (1985) Novel partitioning of DNA cleavage sites for Drosophila topoisomerase II. Cell 40: 933-941.

Vielkind, J. R. (1992) Medaka and zebrafish: ideal as transient and stable transgenic systems. In: Transgenic fish. C. L. Hew and G. L. Fletcher (eds.) World Scientific Publishing Co. pp. 72-91.

Wagner, G.G. and McKeown, B.A. (1986) Development of a salmon growth hormone radioimmunoassay. Gen. Comp. Endo. 62: 452-458.

Yoon, S.J., Hallerman, E.M., Gross, M., Liu, Z., Schneider, J.F., Faras, A.J., Hackett, P.B., Kapuscinski, A.R. and Guise, K.S. (1990) Transfer of the gene for neomycin resistance into goldfish, *Carassius auratus*. Aquaculture, 85: 21-33.

Zhang, P., Hayat, M., Joice, C., Gonzalez-Villasensor, L.I., Lin, C.M., Dunham, R.A., Chen, T.T. and Powers, D.A. (1990) Gene transfer, expression and inheritance of pRSV-rainbow trout-GH-cDNA in the common carp, *Cyprinus Carpio* (Linnaeus). Mol. Reprod. Dev. 25: 3-13.

Zhu, Z., Li, G., He, L. and Chen, S. (1985) Novel gene transfer into the fertilized eggs of goldfish (*Carassius auratus* L. 1758). Z. Angew. Ichthyol. 1: 31-34.

Zhu, Z., Xul, K., Lil G., Xie, Y. and He, L. (1986) Biological effects of human growth hormone gene microinjected into the fertilized eggs of loach *Misgurnus anguillicaudatus* (Cantor). Kexue Tongbao 31: 988-990.

Zhu, Z. (1992) Generation of fast growing transgenic fish, methods and mechanisms. In: Transgenic fish. C. L. Hew and G. L. Fletcher (eds.) World Scientific Publishing Co., Singapore, pp. 92-119.

CHAPTER 8

Transfer of Genes Encoding Neomycin Resistance, Chloramphenicol Acetyl Transferase and Growth Hormone into Goldfish and Northern Pike

Kevin S. Guise, Perry B. Hackett and Anthony J. Faras

ABSTRACT

Gene transfer of the neomycin resistance gene into goldfish was used to study the applicability of this selectable marker (G-418) to piscine systems. Mosaicism was studied in the injected G_0 generation of injected goldfish. The chloramphenicol acetyl transferase gene with a Rous sarcoma virus LTR (RSV) promoter element was injected into goldfish eggs. Fourteen of twenty analyzed fish showed evidence of transfer, with muscle tissue showing most frequent transgene presence and expression. Northern pike eggs were injected with a bovine growth hormone/RSV LTR construct, and preliminary data shows transfer and expression of the bovine gene in a fish system.

INTRODUCTION

Classical systems of genetic selection have led slowly to the development of strains of several commercially important species of fish that have improved growth characteristics. While heavy emphasis has been placed on the rainbow trout (*Salmo gairdneri* , now *Oncorhynchus mykiss*), many species have been virtually ignored in terms of such selection processes. Progress in the manipulation of growth characteristics via selection is a slow process, with rapid enhancement in early generations followed by a slowing of the rate of improvement as the natural variation is "exhausted" by optimization of the choice of possible alleles at each relevant locus. Effects of single genes on growth are difficult to follow during such selection because of the selective pressure placed on all the genes that participate in determining the selected parameter--be it feeding efficiency, growth rate, or age at maturity. The effect of single genes may be assayed by the injection, usually intramuscular, of the product of a growth-related gene, for example, growth hormone (GH). Many attempts have been made to improve the growth rate of fish by treatment with pituitary extracts or purified bovine, ovine, porcine, or piscine GH, and all have produced increased rates of growth--some of which have been truly dramatic. Beginning with the studies of Pickford et al., (1948, 1954), it has been shown that hypophysectomized fish (killifish, *Fundulus heteroclitus*) responded to injections of piscine or bovine GH. Higgs et al., (1975) showed a 30-66% increase in weight over control values at 8 weeks of age in coho salmon (*Oncorhynchus kistch*) after injections of bovine GH or implantation of pellets of GH. Komourdjian et al., (1978) demonstrated the effects of porcine GH on rainbow trout, and Adelman (1977, 1982) reproduced the results using carp GH and bovine GH in carp (*Cyprinus carpio*). The work of Clarke et al., (1977) showed similar effects on *Tilapia mossambica* and *Oncorhynchus nerka* (sockeye salmon) with injection of either Tilapia and bovine GH, and additionally showed that increased levels of GH facilitated the adaptation of salmon to salt water. Komourdjian et al., (1976) confirmed this observation in young Atlantic salmon (*Salmo salar*) treated with porcine GH. In addition to the substantially increased rates of growth, improved feeding efficiency was also noted by Markert et al., (1977) in coho salmon, where injections of bovine GH significantly increased both food intake and protein conversion. Recombinant rainbow trout GH has been shown to produce the same growth effect in rainbow trout (Agellon and Chen, 1986), as have recombinant chicken and bovine GH in Pacific salmon (*Oncorhynchus kisutch*) (Gill et al., 1985). Purified bovine GH has produced not only significant growth, but record-breaking growth, in grass pickerel (*Esox americanus vermiculatus*) in a study by Weatherly and Gill (1978). Grass pickerel in excess of the North American record length were produced through administration of bovine GH. In this latter study, the potential for rapid growth appeared to be correlated with the ability to recruit new muscle fibers into the growing axial muscle mass. All of the above studies involved injection of GH, a process that is both laborious and risky because of the need for repeated injections, and one that introduces a certain amount of stress which may cloud assessment of the effects of the injected hormone. Feeding studies utilizing GH as a feed supplement have shown little success, and only one report (Degani and Gallagher, 1985) has indicated increased growth and food conversion in American elvers (*Anguilla rostrata*). No method for the mass administration of growth factors has yet proven to be economically feasible, nor have any methods shown great promise as techniques for analysis of the effects on growth of products of various growth-promoting genes. Indeed, to date, little work has been devoted to the examination of other growth-promoting peptide hormones or releasing factors by injection or feed supplementation--only GH has been subjected to extensive testing.

Molecular genetics and gene transfer have the potential for introduction of significant new sources of variation within a species, through the introduction of extra copies of cloned genes into the genome and germ line of that species. Since extra copies of a gene generally result in increased levels of the encoded enzyme or protein, the effect on the organism into which these genes have been introduced can be dramatic. The classic experiment in mice by Palmiter et al. (1982) showed a doubling of the mature size in transgenic mice that contained extra copies of genes for either rat or human GH. Since the genes were incorporated into the germ line, this trait was passed in true Mendelian fashion to the offspring, producing a strain of larger mice. Faster growing mice have little enhanced economic value, but increased growth rates in fish are of significant economic interest.

The development of gene transfer in fish by microinjection has mimicked the progress in mice. Zhu et al. (1985) first reported the transfer of a human GH construct into goldfish (*Carassius auratus*) and subsequently the expression, transmission, and significantly enhanced growth rates in the transgenic fish have been reported (Maclean et al., 1987b; see also Zhu, 1992, Chapter 6, this monograph). Early similar transfers of GH genes have been accomplished by Chourrout et al. (1986) in rainbow trout, Zhu et al. (1986) in loach, Dunham et al. (1987) in channel catfish (*Ictalurus punctatus*), McEvoy et al. (1988) in Atlantic salmon, Maclean et al. (1987a) and Powers et al., (1991) in rainbow trout, and by our group in northern pike (*Esox lucius*)(Schneider et al., manuscript in preparation). Expression of genes has been a problem, with few fish expressing the transferred genes when mammalian promoter/enhancer elements are used. Very few heterologous promoter/enhancer elements have been tested as to their utility in fish, namely, RSV (Powers et al., 1991; Yoon et al., 1990) (also Stuart, personal communication), mouse metallothionein (Zhu et al., 1985, Chourrout et al., 1986, McEvoy et al., 1988), and SV40 (Simian Virus 40) (Stuart et al., 1988), with only RSV showing significant promise. In our group, a search is ongoing for appropriate piscine derived promoters, with much of our early work concerning the carp (*Cyprinus carpio*) ß-actin promoter (Liu et al., 1989, in press).

Most gene transfer experiments in fish to date have followed the design of the early mouse work, utilizing in recombinant constructs the mouse metallothionein promoter to regulate the expression of a protein coding sequence. Activities of other promoter sequences have not been demonstrated in transgenic fishes. Retroviral long terminal repeat elements would seem likely to promote the expression of introduced DNA coding sequences because their expression is independent of the location of genomic integration, and in many cases is not tissue or species-specific. While piscine retroviruses have not been well characterized, the key functional domains of the promoter region of the avian Rous sarcoma virus (RSV) have been identified (Luciw et al., 1983) and subcloned (Peterson et al., 1984), enabling their use in the construction of expression vectors.

To date, all gene transfers in fish have been achieved through microinjection of individual eggs, in mammalian cells, however, mass gene transfer has been carried out through calcium phosphate precipitation (Graham and van der Eb, 1973), electroporation (Neumann et al., 1982), or lipofection (Felgner et al., 1987). Such mass transfer schemes generally include a means of selecting transformed cells on the basis of some phenotype conferred through the integration and expression of the introduced DNA construct--often novel drug resistance. No such selection scheme has been demonstrated to date in fish. Demonstration of a protocol for selection of transformants would be a practical contribution toward the development of effective mass gene transfer protocols in fish. A selection regime has been demonstrated (Southern and Berg, 1982) for transformation of mammalian cells with a bacterial gene, neo, conferring resistance to the neomycin analog G-418. The possibility of whole animal selection on the basis of introduced G-418 resistance has been demonstrated in Drosophila (Steller and Pirrotta, 1985). A collaborative group of four laboratories at the University of Minnesota has established the Minnesota Transgenic Fish Group (MTFG) for the study of transgenic processes in fish and to use these processes for improvement of economic traits in fish. These laboratories are those of Dr. Kevin Guise, Animal Science; Dr. Anne Kapuscinski, Fisheries; Dr. Perry Hackett, Genetics; and Dr. Anthony Faras, Human Genetics. The following report outlines some of the studies completed and ongoing by the MTFG.

Our group has studied the transfer into goldfish of a construct bearing the structural gene for neomycin resistance, neo, controlled by the RSV promoter. The objectives of this study were:

1. to evaluate the ability of the RSV promoter to regulate expression of a foreign gene in transgenic fish,
2. to evaluate the usefulness of neomycin resistance as a selectable marker for identification of transgenic individuals expressing novel gene products.

We have also studied the incidence of mosaicism in goldfish using the chloramphenicol acetyl transferase (CAT) gene with a RSV promoter, seeking:

1. to determine the frequency and magnitude of mosaicism in the injected generation,

2. to study the possibility of enhanced expression of a transgene in specific tissues through action of the RSV promoter.

Finally, we have used the RSV promoter in conjunction with a bovine growth hormone gene transferred by microinjection into northern pike to begin an examination of:

1. the ability of a RSV promoter to express constitutively bovine growth hormone (bGH),
2. the potential enhancement of growth rate in northern pike due to the expression of the heterologous bGH.

MATERIALS AND METHODS

Goldfish Embryo Production

Spontaneous ovulation of goldfish was carried out using the methods of Stacy et al (1979). Sexually mature fish were maintained under a long photoperiod regime (16 h light:8 h dark). On day one, breeding fish were transferred from holding tanks to breeding aquaria at 15 ±1 °C. Aquaria were supplied with floating artificial plants, and the water temperature was increased overnight to 21 ±1 °C. Spontaneous ovulation usually occurred during the last half of the dark phase on day three. If no ovulation occurred on day three, the fish were injected intraperitoneally with 2 mg/kg body weight of carp pituitary gland extract. Approximately 10 hours later, eggs and sperm were collected separately by manual stripping of the fish.

Fertilization of goldfish eggs was carried out by the addition of milt followed by the addition of well water. Five minutes after fertilization, goldfish eggs were dechorionated by a four-to-six minute incubation in 0.25% trypsin in Holtfeter's solution (Grand et al., 1941). Dechorionation was stopped by addition of 5% fetal bovine serum in Holtfreter's solution. Dechorionated eggs were washed several times in Holtfreter's solution and transferred to Holtfreter's solution on an agar Petri dish.

Northern Pike Embryo Production

Eggs and sperm were collected from wild caught northern pike (*Esox lucius*). Fertilization at room temperature was followed by water hardening in well water. Eggs were injected with chorion intact, since no known enzyme treatment has been found to remove northern pike chorions (Grand et al., 1941).

Plasmid Construction

The plasmid pRSVneo contains the neo structural gene, whose product (aminoglycoside 3'-phosphotransferase II) confers resistance to the neomycin analog G-418, under the control of the RSV promoter (Yoon et al., 1990) The neo structural gene is under the control of the functional domains of the promoter region of the avian Rous sarcoma virus (RSV) (Luciw et al., 1983; Peterson et al., 1984). The pRSVcat construct incorporates the coding for the chloramphenicol acetyltransferase protein fused to the 5' long terminal repeat of the Rous sarcoma virus (Gorman et al., 1982). Its function was verified by observation of antibiotic resistance in transformed *E. coli* cells. The pRSVbGH construct is a simple replacement of the neo gene of pRSVneo with a bGH cDNA.

Microinjection

The pRSVneo was linearized with *Kpn*I (which cuts in the polylinker of the plasmid), redissolved in 88 mM NaCl, 10 mM Tris HCl, pH 7.6 to a final concentration of 25 μl/ml. The plasmids pRSVcat and pRSVbGH were used in supercoiled form. Borosilicate glass microinjection needles with an inner tip diameter of approximately 2 μm were prepared by use of a vertical pipette puller (David Kopf Instruments, Model 700-C). The needles were filled with plasmid and blue dye (FD&C Blue #1, McCormick and Co.) solution and microinjection was carried out using a Brinkman MM33 micromanipulator. Approximately 2 nl of plasmid solution was injected, with volume controlled by timing the insertion/withdrawal of the injection needle at constant fluid flow rate. DNA was injected into the center of the germinal disc of the fish eggs prior to first cleavage.

Probe Labelling

To derive a probe for Southern and DNA dot-blot analysis, a 2.3 kb *Bam*HI-*Hind*III restriction fragment of pRSVneo was isolated by standard techniques, and labelled by the procedure of Feinberg and Vogelstein (1983) to 0.5-2 x 10^9 cpm/μg. DNA probes were used without purification from unincorporated nucleotides. This 2.3 kb *Bam*HI-*Hind*III fragment including the neo gene was also subcloned into pTZ1BR (Pharmacia) under the control of the phage T7 promoter to produce a transcript complementary to neo mRNA. One μg of linear DNA was transcribed using ^{32}P-UTP following the method of Schenborn and Mierendorf (1985) to produce probes used in

RNA dot-blot analyses of goldfish. Similar fragments containing the entire RSVcat and RSVbGH inserts were prepared .

DNA Analysis-- Neo gene

Genomic DNA was isolated from 1-2 month old goldfish which had been bisected and stored at -90 °C Posterior halves of fish were individually homogenized in three volumes of 0.2 M Tris HCl (pH 8.0), 0.1 M EDTA, 0.5% (w/v) sodium dodecyl sulfate (SDS) buffer with 100 µg/ml proteinase K (Boehringer-Mannheim) using a Dounce homogenizer (Wheaton). High molecular weight DNA was extracted following standard methods (Maniatis et al., 1982).

DNA dot-blot analysis (Maniatis et al., 1982) was used to detect the presence of the neo gene within the genome of goldfish. Five µg of genomic DNA was denatured in 0.4 M NaOH and, following addition of an equal volume of 2 M ammonium acetate, was spotted onto a nitrocellulose filter using a Hybri-dot system (Bethesda Research Laboratories). Hybridization solution was 50% deionized formamide, 5x SSC (1x SSC = 0.15 M NaCl, 15 mM Na citrate, pH 7.2), 5x Denhardt's solution (1x Denhardt's = 0.02% Ficoll, 0.02% bovine serum albumin, 0.02% polyvinylpyrolidone), 50 mM sodium pyrophosphate (pH 6.5), 0.1% SDS, 0.1 µg/ml denatured calf thymus DNA (Sigma), and 0.1 µg/ml yeast tRNA (Boehringer-Mannheim) to which 1 x 10^6 cpm/ml of ^{32}P-labeled probe was added. Highest stringency was in 0.2x SSC, 0.1% SDS (Maniatis et al., 1982). Southern blot analysis (Southern, 1975) was carried out on 10 µg of goldfish genomic DNA completely digested with *Kpn*I, *Bam*HI, or *Sst*I (each of which cuts the pRSVneo plasmid only once). The Southern blot was probed using the same hybridization conditions as the dot-blot analysis.

Persistence of DNA Construct

High molecular weight DNA was purified (Sambrook et al., 1989) from the following tissues of the goldfish: muscle, skin, spleen, liver, and gonad. Persistence of the DNA construct in the tissues of the host was assayed by Southern hybridization (Southern, 1975) of EcoR1 digests using the 2533 bp EcoR1 restriction fragment of the RSVcat construct as the hybridization probe to detect presence of the transgene. The probe was labelled with [^{32}P] to a specific activity of 2 x 10^9 cpm/µg by the random primer method (Sambrook et al., 1989), as per manufacturer's (Pharmacia) protocols. Autoradiography was carried out for 7 days at -70°C, using Kodak XAR-5 film and intensifying screens.

RNA Analysis-- Neo gene

Anterior halves of fish stored at -90°C were homogenized using a polytron (Kinematica) in a solution containing 4 M guanidinium isothiocyanate, 0.1 M beta-mercaptoethanol, 0.5% sarkosyl, 5 mM sodium citrate, pH 7.0. Total RNA was isolated following the methods of Maniatis et al., (1982) or Davis et al., (1986). RNA dot-blot analysis (White and Bancroft, 1982) was used to test for neo gene expression.

G-418 Selection

For the drug selection screenings, solutions of G-418 in static water aquaria were aerated to achieve initial saturation levels of dissolved oxygen. Dissolved oxygen was maintained above 60% saturation by periodic aeration. Temperature was maintained at 20 ±1 °C. The pRSVneo-injected or non-injected control goldfish at a population density of 12 g/l were exposed separately to 4 mg/ml or 8 mg/ml solutions of G-418. Mortality of fish was monitored every four hours for the first 24 hours and every 12 hours afterward. Dead individuals were removed upon observation and stored in liquid nitrogen. Drug selection was continued until 50% of the control fish in 4 mg/ml G-418, or until 90% of those in 8 mg/ml, had died. Surviving fish were transferred into fresh well water upon termination of the test.

CAT Activity

Quantification of CAT enzyme activity followed Gorman (1982a,b). Briefly, 0.3 uCi [^{14}C]-chloramphenicol (59.5 mCi/mM specific activity) was incubated overnight (18 hr) with tissue homogenate. The reaction mix was extracted with ethyl acetate, and the reaction products were separated using thin layer chromatography. An autoradiograph was used to localize and remove spots on the chromatogram corresponding to acetylated and unacetylated [^{14}C]-chloramphenicol which was then quantified through liquid scintillation counting. The concentration of protein in tissue homogenates was quantified by using a Bio-Rad protein assay (BioRad, Richmond, CA) based upon the protein-dye binding protocol of Bradford (1976). Specific activity of CAT in a given tissue sample was calculated as the number of counts of acetylated [^{14}C]-chloramphenicol divided by the number of micrograms of total protein in the assay. The CAT activity associated with a given tissue sample was considered to be above background if it was twice the highest value observed among control samples drawn from the respective tissues.

Analysis of Bovine Growth Hormone Gene Injected Northern Pike

Co-transfer of pRSVcat and pRSVbGH was performed into northern pike eggs. Bovine growth hormone radioimmunoassay (RIA) was performed by the method of Wheaton (1986) on serum sampled from the dorsal aorta of the RSVbGH injected northern pike, using a rabbit anti-bGH produced by our group. Southern blots were assayed for both pRSVcat and pRSVbGH presence. Fin tissue was asssayed, as were blood samples, for CAT activity as above.

RESULTS

Molecular Analysis of pRSVneo Injected Fish

The RSVneo construct was successfully introduced into goldfish using standard microinjection techniques. The survival rate for groups of microinjected fish has ranged from 10% in early experiments to close to 50% currently, compared with survival rates of above 90% in uninjected control groups.

To determine what percentage of the surviving fish were transgenic, i.e., bearing the RSVneo construct integrated within their genome, DNA isolated from the fish was analyzed by DNA dot-blots. Neither of the control fish showed positive hybridization to the neo probe. Four out of 59 pRSVneo-injected fish showed hybridizations of intensities equivalent to approximately one copy of the gene per genome. All dot-blot analyses were repeated, and yielded identical results. Thus, roughly 8% of the goldfish surviving through one month of age carry the transferred construct in at least some of their tissues (Yoon et al., 1990).

To examine the mode of genomic integration, the DNA of putative transgenic fish was further examined in Southern blots. The neo probe detected a 5.3 kb band with mobility equal to that of the linearized plasmid in all four test individuals which may be evidence of circularization and persistence of the introduced genetic construct. Higher and lower molecular weight bands observed in one individual were interpreted as being possible junction fragments indicative of genomic integration of the introduced construct. Results from Southern blots of *Sst*I - and *Kpn*I-digested DNA agreed with the results with *Bam*HI-digested DNA (Yoon et al., 1990).

To determine to what extent the transferred neo gene was expressed, dot-blots of total RNA prepared from the fish were probed for evidence of the neo-specific mRNA. Expression of the neo gene was evident on the RNA dot-blot of two of the four putative transgenic fish, indicating that about half of the fish that show evidence of transgene DNA also give evidence of expression (Yoon et al., 1990).

G-418 Selection of pRSVneo Injected Fish

Forty pRSVneo-injected fish were exposed to 4 mg/ml G-418. After 20 had died, the remaining fish were transferred to fresh well water. All fish died within seven days after the test, presumably due to stress attributable to the drug exposure. One fish, which survived the drug selection trial but later died of fungal infection, showed a hybridization signal with an intensity of roughly one gene copy per genomic equivalent. DNA from groups of fish sensitive to G-418 showed no significant hybridization signal (Yoon et al., 1990).

In a second G-418 trial, sixty-three pRSVneo-injected fish were exposed to 8 mg/ml G-418. After 90% of the control fish exposed to G-418 in a separate vessel had died, the nine surviving test fish were transferred to fresh well water. The test individuals died within three days after the test. DNA was extracted from these fish and analyzed by dot-blots. DNA from three of these individuals showed hybridizations with an intensity of less than one gene copy per genomic equivalent. This may be evidence of chimerism in these individuals with respect to the introduced pRSVneo construct (Yoon et al., 1990).

Background CAT Activities in Goldfish Tissues

Despite earlier reports that eukaryotic cells exhibit no endogenous CAT activity (Gorman et al., 1982a, 1982b) we did observe a low-level background of acetylated chloramphenicol in control fish tissue homogenates. Goldfish cells may indeed lack specific CAT activity, but with the long incubations (18 hr) used in this study, activities presumably of non-specific acetyltransferases reached observable thresholds. Such background CAT activities were generally less than 10 counts/μg protein. Higher background CAT activities were observed in skin and in gonad and muscle.

Persistence of pRSVcat

Screening of Southern blots of DNA samples prepared from tissues of 20 test goldfish indicated the persistence of the pRSVcat construct within certain tissues. The CAT gene probe hybridized to high molecular weight DNA samples from 11 of the 20 test individuals. Hybridization alone could not rule out the possibility of extrachromosomal persistence of the introduced DNA construct (Maclean et al., 1987b).

CAT Activity in Goldfish Tissues

Tissues from six control and 20 fish hatched from microinjected eggs were assayed for CAT activity. Fourteen test individuals bore at least one tissue exhibiting CAT activity of a level at least twice that observed in the corresponding tissue of any control fish. The distribution of elevated levels of CAT activity suggested some tissue specificity of expression.

DISCUSSION

The results reported here showed successful transfer of the pRSVneo construct into goldfish, *Carassius auratus*. Integration of the transferred construct was strongly suggested. Some of the transgenic individuals exhibited hybridization phenotypes consistent with single genomic integration sites. The observation of junction fragments indicated possible genomic integration of the microinjected construct. The multi-banded hybridization pattern of function fragments thus exhibited may be indicative of multiple genomic integration sites, presumably due to delayed genomic integration with subsequent mosaicism for the neo . Expression of specific RNA to the neo gene was detected in about half of the transgenic animals.

Table 1. Tissue distribution of CAT activity in injected goldfish.

Tissue	Muscle	Skin	Brain	Pooled Internal Organs
CAT Activity Detected	70 %	5%	30%	10%
CAT DNA Presence Detected	60%	40%	10%	25%

Over 1200 pRSVcat and pRSVbGH injected pike were analyzed at about 7 months after hatching. Approximate transfer and expression efficiencies are as follows :

Table 2. Preliminary Northern Pike Analysis Summary.

Total Fish Analyzed	Containing Transgene	Positive bGH RIA (serum)	Positive for CAT expression
1269	5%	4%	12%

Identification of promoters active in fish is an important practical aspect of developing gene transfer as a means of genetic improvement. Adding to previous reports of metallothionein promoter activity in fish (Maclean et al., 1987b), our data indicated the activity of the RSV promoter in the genome of goldfish. Definitive demonstration of correct function of the RSV promoter in goldfish would require evidence of proper initiation of mRNA transcription (through S1 nuclease mapping or primer extension), production of full-length mRNA transcripts (Northern blot), and translation of mRNA into its protein product (radioimmunoassay). Activity of the RSV promoter also has been inferred in northern pike (Schneider, manuscript in preparation) and in zebrafish (G.W. Stuart, University of Oregon at Eugene, personal communication and Liu et. al., submitted).

Our group has been exploring the use of mass transfer techniques to overcome the tedium of injecting the large numbers of eggs needed to generate a phenotypically improved transgenic fish. While mass transfer techniques might prove less efficient than microinjection, the ability to manipulate large numbers of eggs at one time might compensate for the loss in transfer efficiency. A key element necessary to make mass transfer feasible is the development of a selection system to differentiate transgenic from non-transgenic fish without sacrifice of the transgenic fish. One approach to such a selection system would be to distinguish transgenic from non-transgenic individuals on the basis of transferred antibiotic resistance. The neo gene, conferring resistance to G-418, was utilized in this gene transfer study because of its known potential as a selectable marker in a wide range of eukaryotic organisms.

A gene conferring a selectable phenotype would ultimately be utilized as one component of a construct which also bears a gene of economic importance. Thus, selection would identify individuals likely to exhibit economically improved phenotypes. Individuals exhibiting high levels of marker-gene expression would be likely to exhibit high levels of expression of the economically important gene.

Trials based upon neo expression as a selectable marker were inconclusive. Some putatively chimeric individuals did not survive exposure of G-418. Mosaicism for the introduced neo construct would complicate the response of an individual to G-418 selection. If genomic integration of the construct had been delayed beyond the stage of differentiation of cell lineages in the embryo, then only certain tissues of the adult would be transgenic. Thus, the sensitivity or resistance to a given concentration of G-418 of a chimeric individual might be a function of how many tissues and which particular tissues were expressing the introduced gene. The problem of response of chimeric individuals to drug selection may ultimately have to be circumvented by testing the progeny derived from the original transgenic individuals. Mosaic parents with some transgenic germ cells would pass the novel gene to a portion of their progeny. These transgenic progeny would not be chimeric. See Vielkind (1992) for discussion concerning variegated expression of transgenes (Chapter 5, this monograph).

We also detected some individuals which were not demonstrably transgenic that survived exposure to G-418. While there could be a small amount of genetic variance for G-418 sensitivity in these fish, this is largely eliminated as a possibility due to the failure to detect such survivors in the control non-injected population. Alternatively, these individuals may have been chimeric but with transgenic cells at low frequencies. In such cases, dot-blot analysis can fail to identify chimeric individuals, because a single copy of introduced DNA in as few as one-fifth of the cells would be undetectable (Wilkie et al., 1986)· Cases have been noted in transgenic mice where DNA analysis did not reveal incorporation of introduced DNA, but biological activity of the product of the introduced gene was detected (Hammer et al., 1985).

Our concern with the potential problems poised by mosaicism was addressed in the pRSVcat study which sought to determine the level of mosaicism in the injected generation of transgenic fish. The relatively easy assay of CAT expression also allowed the determination of the activity of the RSV promoter. The tissue specificities and levels of gene expression consequent to transfer of genetic constructs driven by the RSV promoter have been characterized in chicken (Howlett et al., 1987, Hippenmeyer et al., 1988) and in mouse (Overbeek et al., 1986). RSV LTR-regulated expression vectors have been introduced into carp (Zhang et al., 1988), goldfish (Yoon et al., 1990), zebrafish and northern pike (Schneider et al., 1989). The tissue specificities and levels of expression promoted by the RSV LTR element have recently been characterized in piscine systems (Hallerman et al., in press).

Following incorporation of a potentially useful promoter element and a marker gene into an expression vector and screening for the transgene product in a recipient species, the tissue specificity and levels of expression produced by the regulatory element can thus be characterized (Jaenisch, 1988). Transgene persistence and levels of expression of the chloramphenicol acetyltransferase (CAT) marker gene under the regulation of the RSV LTR element were examined in several tissues of goldfish hatched from eggs microinjected with the recombinant DNA construct. As may be seen from the results reported above, most of the randomly chosen microinjected fish were mosaic for pRSVcat (14/20 or 70%). Only one fish had transgene DNA or CAT expression at detectable levels in all analyzed tissues.

The pRSVcat data from the mosaicism study may also be subdivided into four groups based on the presence or absence of detectable DNA versus the expression of the CAT transgene. The expression of the CAT gene in the presence of transgene DNA, and the lack of CAT gene expression in the absence of transgene DNA are self explanatory. The lack of CAT gene expression in the presence of transgene DNA may represent simply the persistence of an incomplete transgene. The opposite circumstance, the observation of CAT expression in the absence of detectable transgene DNA is not without precedent and may be explained by the relative sensitivity of CAT versus DNA analyses. Detectable levels of CAT expression might occur in a subset of the cells in a tissue, while a single copy of a plasmid in the same subset could prove undetectable (Wilkie et al., 1986).

In the transgenic goldfish examined in this study, the highest levels of RSV-induced transgene activity were observed in muscle, with lesser levels of activity sometimes observed in brain and internal organs. Others have shown in mammals (Gorman et al., 1982a,b; Overbeek et al., 1986) that the amount of CAT-specific mRNA encoded by pRSVcat correlates with the level of CAT enzyme activity. Assuming that the CAT mRNA and CAT protein are equally stable in all tissues, the expression of the CAT transgene under the regulation of the RSV LTR observed in transgenic goldfish in this study is consistent with previous observations in tissues of adult transgenic mice (Overbeek et al., 1986) and chicken embryos (Howlett et al., 1987; Hippenmeyer et al., 1988) where the RSV LTR is active in muscle cells.

The observed expression of RSV LTR-regulated transgenes in tissues of mesodermal origin also parallels the observation of tumorigenesis in such tissues following RSV infection in chicken (Purchase and Burmester, 1978) and mouse (Svet-Moldavsky, 1958) The molecular basis for this tropism has been linked to a response by the viral regulatory elements to tissue-specific stimuli (Overbeek et al., 1986), the precise nature of which is uncharacterized. Although LTR-induced transgene expression was generally observed in tissues normally transformed by RSV, the observation of ectopic expression is not without precedent, having been observed in mice (Soriano et al., 1986; Small et al., 1986; Nerenberg et al., 1987). Ectopic expression in some individuals has been hypothesized to be a consequence of the site of genomic integration (Soriano et al., 1986, Lang et al.,1987).

The detection of transgene in some but not all tissues of an individual fish suggests probable mosaicism for the pRSVcat construct. Mosaicism for introduced DNA constructs in a wide range of animals (Flytzanis et al., 1985; Wilkie, 1986; Etkin and Pearman, 1987; Stuart et al., 1988; Fletcher and Davies, 1991) has been attributed to a delay in the integration of the transgene construct into host genomic DNA. This was probably the case among certain of the transgenic goldfish examined in this study. The Southern analysis of these fish suggest that much of the DNA detected is either unintegrated or integrated as tandem copies. Delayed integration may also be the cause of the widely disparate gene expression seen in muscle and skin samples from several individuals. In these individuals, a large difference was detected in CAT activities between two separate tissue samples derived from the same tissue of an individual. It is possible that our random sampling detected patches of tissue that were derived from different clonal lineages. It has been noticed that biological activity of introduced constructs has been observed in tail tissues of transgenic mice where analysis of tail DNA revealed no integrant (Hammer et al., 1985). Thus detection of transgenesis in the microinjected (G_O) individual requires assay of multiple tissues to counter the complication of mosaicism, and should entail both analysis for transgene detection (Southern) and gene expression. Only in this manner may a more accurate picture of the degree of success achieved in production of a transgenic animal be obtained.

The preliminary northern pike data indicates similar levels of transfer as seen in the marker genes. Almost 20% of the G_O population (the injected generation) show some evidence of transgene presence or activity in either blood or fin tissue (Schneider et al., in preparation). Growth data is being compiled with indications of an average growth enhancement of about 40% in the injected G_O population (Schneider et al., 1989; Gross et al., in preparation).

SIGNIFICANCE

The elevated level of expression in muscle tissue of CAT transgenes promoted by the RSV LTR element suggests its utility in expression vectors intended for practical genetic engineering applications. For example, fast-growth of transgenic animals might result from elevated levels of expression of growth factors in muscle tissue. This hypothesis is supported by preliminary results from two experiments where RSV-LTR/growth hormone constructs were introduced into fish. Accelerated growth was observed among northern pike bearing an introduced RSV-LTR/bovine growth hormone construct (Schneider et al., 1989) and among common carp bearing an introduced RSV-LTR/rainbow trout growth hormone construct (Zhang et al., 1988).

Since gene transfer by microinjection in fish has been proven effective, the search for fish growth-promoting genes has intensified. The effort to clone and express piscine genes that promote growth in fish is worldwide. GH genes have been cloned from rainbow trout (Agellon and Chen, 1986), gilthead seabream (*Sparus aurata*) (Cavari et al., 1988), chum salmon (*Oncorhynchus keta*) (Sekine et al., 1985) coho salmon (Nicoll et al., 1987, Gonzalez-Villasenor et al., 1988), chinook salmon (Hew et al., 1989), yellowtail (*Seriola quinqueradiata*) (Watahiki et al., 1988), tuna (*Thunnus thunnus*) (Sato et al., 1988), and red seabream (*Pagus major*) (Momota et al., 1988). Chinook salmon prolactin has been cloned by Song et al. (1988) and represents the start of efforts to clone other genes for piscine peptide hormones. One non-growth-regulating gene, the antifreeze protein of winter flounder (*Pseudopleuronectes americanus*), has been cloned and transferred by Fletcher et al., (1988) into Atlantic salmon (See also Fletcher et al., 1992, Chapter 11, this monograph).

Our group is engaged in the construction of piscine expression vector utilizing sequences (promoter and structural gene) derived only from fish. Such a vector should produce transgenic fish more acceptable to the consuming public than transgenic fish with heterologous genes. To this end we have cloned the carp ß-actin gene and have studied the enhancer/promoter elements of the gene for use in such an expression vector (Liu et al., 1990).

These studies point to the practicality and potential of transgenic, growth enhanced fish. Results of early experiments such as ours are only now reaching publication, in part due to the long culture/generation time of the aquaculture species utilized. The next few years will show the firm transmission of these genes to the second generation, and will study the stability of expression. These are mandatory before commercialization is feasible.

ACKNOWLEDGMENTS

This project has received support from a variety of organizations including Minnesota Sea Grant, the Minnesota State Legislature through the Legislative Commission on Minnesota Resources, the Minnesota Experiment Station, the Blandon Foundation, and the Greater Minnesota Corporation. A collaborative group of four laboratories at the University of Minnesota has established the Minnesota Transgenic

Fish Group (MTFG) for the study of gene transfer in fish and to use it for the improvement of economic traits in fish. These laboratories are those of Dr. Kevin Guise , Animal Science; Dr. Anne Kapuscinski, Fisheries; Dr. Perry Hackett, Genetics; and Dr. Anthony Faras, Human Genetics. Many people within MTFG have been instrumental in the success of this work. They include John Schneider, Eric Hallerman, Sung Joo Yoon, Ling He, Mark Gross, Zhanjiang Liu, Boaz Moav and Neomi Moav (Israel), Z. Zhu (People's Republic of China), and C. Hew (Canada).

REFERENCES

Adelman, I.R. (1977) Effect of bovine growth hormone on the growth of carp *(Cyprinus carpio)* and the influences of temperature and photoperiod. J. Fish. Res. Board Can. 34: 509-515.

Adelman, I.R. (1982) Enhancement of growth of common carp by injection of a homogenate of carp pituitary glands. Prog. Fish-Cult. 44: 94-97.

Agellon, L.B. and Chen, T.T. (1986) Rainbow trout growth hormone: Molecular cloning of cDNA and expression in *Escherichia coli..* DNA 5: 463-467.

Bradford, M.M. (1976) A rapid and sensitive method for the quantification of microgram quantities of protein utilizing the principle of protein-dye binding. Anal. Biochem. 72: 248-254.

Cavari, B., Funkenstein, B., Chen, T.T. and Powers, D.A. (1988) Cloning the growth hormone gene from the sea bream *Sparus aurata..* Third Intl. Conf. Genet. Aquacult., Trondheim, Norway, 20-24 June, 1988 (abstract).

Chourrout, D., Guyomard, R. and Houdebine, L.M. (1986) High efficiency gene transfer in rainbow trout (*Salmo gairdneri* Rich.) by microinjection into egg cytoplasm. Aquaculture 51: 1443-1150.

Clarke, W.C. , Farmer, S. and Hartwell, K. (1977) Effect of teleost pituitary growth hormone on growth of *Tilapia mossambica* and on the growth and seawater adaptation of sockeye salmon (*Oncorhynchus nerka*). Gen. Comp. Endocrinol. 33: 174-178.

Davis, L.G., Dibner, M.D. and Battey, J.E.(1986) Basic Methods in Molecular Biology. Elsevier, NY, pp. 136-138 .

Degani, G.and Gallagher, M.L. (1985) Effects of dietary 17a-methyltestosterone and bovine growth hormone on growth and food conversion of slow- and normally-growing American elvers (*Anguilla rostrata*). Can. J. Fish. Aquat. Sci. 42: 185-189.

Dunham, R.A., Eash, J., Askins, J. and Towns, T.M. (1987) Transfer of the metallothionein-human growth hormone fusion gene into channel catfish. Trans. Am. Fish. Soc. 116: 87-91.

Etkin, L.D. and Pearman, B. (1987) Distribution, expression, and germ-line transmission of exogenous DNA sequences following microinjection into *Xenopus laevis* eggs. Development 99: 15-23.

Feinberg, A.P. and Vogelstein, B. (1983) A technique for radiolabelling DNA restriction fragments to high specific activity. Anal. Biochem. 132: 6-13.

Felgner, P.L., Gadek, T.R., Holm, M., Roman, R., Chan, H.W., Wenz, M., Northrup, J.P., Ringold, G.M. and Danielsen, M. (1987) Lipofection: A highly efficient, lipid mediated DNA transfection procedure. Proc. Natl. Acad. Sci., U.S.A. 84: 7413-7417.

Fletcher, G.L., Shears, M.A., King, M.J., Davies, P.L. and Hew, C.L. (1988) Evidence for antifreeze protein gene transfer in Atlantic salmon (*Salmo salar*). Can. J. Fish. Aquat. Sci. 45: 352-357.

Fletcher, G. L. and Davies, P. L. (1991) Transgenic fish for aquaculture. In: Genetic engineering, principles and methods. J. K. Setlow (ed.) Plenum Press, New York, 331-370.

Fletcher, G. L., Davies, P. L. and Hew, C. L. (1992) Genetic engineering of freeze-resistant Atlantic salmon. In: Transgenic fish. G. L. Fletcher and C. L. Hew (eds.). World Scientific Publishing Co., Singapore. pp. 190-208.

Flytzanis, C.N., McMahon, A.P., Hough-Evans, B.R., Katula, K.S., Britten, R.J. and Davidson, E.H. (1985) Persistance and integration of cloned DNA in post-embryonic sea urchins. Dev. Biol. 108: 431-442.

Gill, J.A., Sumpter, J.P., Donaldson, E.M., Dye, H.M., Souza, L., Berg, T., Wypych, J. and Langley K. (1985) Recombinant chicken and bovine growth hormones accelerate growth in aquacultured juvenile pacific salmon (*Oncorhynchus kisutch*). Bio/Technology 3: 643-646.

Gonzalez-Villasenor, L.I., Zhang, P., Chen, T.T. and Powers, D.A. (1988) Molecular cloning and sequencing of coho salmon growth hormone cDNA. Gene 65: 239-245.

Gorman, C.M., Moffat, L.M. and Howard, B.H. (1982) Recombinant genomes that express chloramphenicol acetyltransferase in mammalian cells. Mol. Cell. Biol., 2: 1044-1051.

Gorman, C.M., Merlino, G.T., Willingham, M.C., Pastan, I. and Howard, B.H. (1982) The Rous sarcoma virus long terminal repeat is a strong promoter when introduced into a variety of eukaryotic cells by DNA-mediated transfection, Proc. Natl. Acad. Sci. USA 79: 6777-6781.

Graham, F.L. and van der Eb, A. (1973) A new technique for the assay of infectivity of human adenovirus 5 DNA. Virology 52: 456-467.

Grand, C.G., Gordon, M. and Cameron, G. (1941) Neoplasma studies: Cell types in tissue culture of fish melanotic tumors compared to mammalian melanomas. Cancer Res. 1: 660-666.

Hallerman, E.M., Schneider, J.F., Gross,M., Liu, Z., Yoon, S.J., Hackett, P.B., Faras, A.J., Kapuscinski, A.R. and Guise, K.S. (1990) Gene expression promoted by the RSV long terminal repeat element in transgenic goldfish. Animal Biotechnology. 1: 79-93.

Hammer, R.E., Brinster, R.L. and Palmiter, R.D. (1985) Use of gene transfer to increase animal growth, Cold Spring Harbor Symp. Quant. Biol. 50: 379-388.

Hew, C. L., Trinh, K. Y., Du, S. J. and Song, S. (1989) Molecular cloning and expression of salmon pituitary hormones. Fish Physiol. Biochem. 7: 375-380.

Higgs, D.A., Donaldson, E.M., Dye, H.M. and McBride, J. R. (1975) A preliminary investigation of the effect of bovine growth hormone on growth and muscle composition of coho salmon (*Oncorhynchus kisutch*). Gen. Comp. Endocrinol. 27: 240-253.

Hippenmeyer, P.J., Krivi, G.G. and Highkin, M.K. (1988) Transfer and expression of the bacterial NPT-II gene in chick embryos using a Schmidt-Ruppin retrovirus vector. Nucleic Acids Res. 16: 7619-7632.

Howlett, A.R., Cullen, B., Hertle, M. and Bissell, M.J. (1987) Tissue tropism and temporal expression of Rous-sarcoma virus in embryonic avian limb in ovo. Oncogene Res. 1: 255-263.

Jaenisch, R. (1988) Transgenic Animals. Science 240: 1468-1474.

Komourdjian, M.P., Saunders, R.L. and Fenwick, J.C. (1976) The effect of porcine somatotropin on growth and survival in seawater of Atlantic salmon (*Salmo salar*). Can. J. Zool. 54: 531-535.

Komourdjian, M.P., Burton, M.P. and Idler, D.R. (1978) Growth of rainbow trout, *Salmo gairdneri*, after hypophysectomy and somatotropin therapy. Gen. Comp. Endocrinol. 34: 158-162.

Lang, R.A., Metcalf, D., Cuthbertson, R.A., Lyons, I., Stanley, E., Kelso, A., Kannourakis, G., Williamson, D.J., Klintworth, G.K., Gonda, T.J. and Dunn, A.R. (1987) Transgenic mice expressing a hematopoetic growth factor (GM-CSF) develop accumulations of macrophages, blindness, and a fatal syndrome of tissue damage. Cell 51: 675-686.

Liu,Z., Zhu, Z., Faras, A.J., Guise, K.S., Kapuscinski, A.R. and Hackett, P.B. (1989) Sequence of the grass carp b-actin gene. Nucleic Acids Res. 17: 5850.

Liu,Z., Zhu, Z., Roberg, K., Faras, A.J., Guise, K.S., Kapuscinski, A.R. and Hackett, P.B. (1990) Isolation and characterization of the b-actin gene of carp (*Cyprinus carpio*) DNA sequence. J. DNA Sequence and mapping. 1: 125-136.

Luciw, P.A., Bishop, J.M., Varmus, H.E. and Capecchi, M.R. (1983) Location and function of retroviral and SV40 sequences that enhance biochemical transformation after microinjection of DNA. Cell 33: 705-716.

Maclean, N., Penman, D., and Talwar, S. (1987a) Introduction of novel genes into the rainbow trout, Pages 325-328 in K. Tiews, ed. *Selection, Hybridization, and Genetic Engineering in Aquaculture*. Verlag, H. Heenemann GmbH, Berlin.

Maclean, N., Penman, D. and Zhu, Z. (1987b) Introduction of novel genes into fish. Bio/Technology, 5: 257-261.

Maniatis, T., Fritsch, E.F. and Sambrook, J. (1982) Molecular Cloning: A Laboratory Manual. Cold Spring Harbor Laboratory, Cold Spring Harbor, NY.

Markert, J.R., Higgs, D.A., Dye, H.M. and MacQuarrie, D.W. (1977) Influence of bovine growth hormone on growth rate, appetite, and food conversion in yearling coho salmon (*Oncorhynchus kisutch*) fed two diets of different composition. Can. J. Zool. 55: 74-83.

McEvoy, T., Stack, M., Keane, B., Barry, T., Sreenan, J. and Gannon, F. (1988) The expression of a foreign gene in salmon embryos. Aquaculture 68: 27-37.

Momota, H., Kosugi, R., Hiramatsu, H., Ohgai, H., Hara, A. and Ishioka, H. (1988) Nucleotide sequence of cDNA encoding the pregrowth hormone of red sea bream (*Pagus major*). Nuc. Acids Res. 16: 3107.

Nerenberg, M., Hinrichs, S.H., Reynolds, R.K., Khoury, G. and Jay, G. (1987) The tat gene of human t-lymphocytic virus type I induces mesenchymal tumors in transgenic mice. Science. 237: 1324-1329.

Neumann, E. , Schafer-Ridder, M., Wang, Y. and Hofschneider, P.H. (1982) Gene transfer into mouse myloma cells by electroporation in high electric fields. EMBO J. 1: 841-845.

Nicoll, C.S., Steiny, S.S., King, D.S., Nichioka, R.S., Mayer, G.L., Ederhardt, N.L., Baxter, J.D., Yamanaka, M.K., Miller, J.A., Seilhamer, J.J., Schilling, J.W. and Johnson, L.K. (1987) The primary structure of coho salmon growth hormone and its cDNA. Gen. Comp. Endocrinol. 68: 387-399.

Overbeek, P.A., Lai S.P., Van Quill, K.R. and Westphal, H. (1986) Tissue-specific expression in transgenic mice of a fused gene containing RSV terminal sequences. Science 231: 1574-1577.

Palmiter, R.D., Brinster, R.L., Hammer, R.E., Trumbauer, M.E., Rosenfeld, M.G., Brinberg, N.C. and Evans, R.M. (1982) Dramatic growth of mice that develop from eggs microinjected with metallothionein-growth hormone fusion genes. Nature 300: 611-615.

Peterson, R.B., Hansel, C.H. and Hackett, P.B. (1984) Identification of a ribosome-binding site for a leader peptide encoded by Rous sarcoma virus RNA. J. Virol. 51: 722-729.

Pickford, G.E. and Thompson, E.F. (1948) The effects of purified mammalian growth hormone on the killifish *Fundulus heteroclitis* (Linn). J. Exp. Zool. 109: 367.

Pickford, G.E. (1954). The response of hypophysectomized male killifish to purified fish growth hormone, as compared with the response to purified beef hormone. Endocrinology 55: 274-287.

Powers, D.A., Gonzales-Villasenor, L.I., Zhang, P., Chen T.T. and Dunham, R.A. (1991) Transgenic studies on fish: transfer, expression, and inheritance. In: Transgenic animals. First and Haseltime (eds.). Butterworth-Heinemann, 307-324.

Purchase, H.G. and Burmester, B.R. (1978) Leukosis/sarcoma group, In: Diseases of Poultry. Hofstad, Calvek, Helmbolt, Reid, and Yoder (eds.) Iowa State University Press, Ames, IA, pp. 418-468.

Sambrook, J., Fritsch, E.F. and Maniatis, T. (1989) *Molecular Cloning: A Laboratory Manual, Second Edition,* Cold Spring Harbor Laboratory, Cold Spring Harbor, NY.

Sato, N., Watanabe, K., Murata, K., Sakaguchi, M., Kariya, Y., Kimura, S., Nonaka, M. and Kimura, A. (1988) Molecular cloning and nucleotide sequence of tuna growth hormone cDNA. Biochim. Biophys. Acta 949: 35-42.

Schenborn, E.T. and Mierendorf, R.C., Jr. (1985) A novel transcription property of Sp6 and T7 RNA polymerases: Dependence on template structure. Nucleic Acids Res. 13: 6223-6236.

Schneider, J.F., Hallerman, E.M., Yoon, S.J., He, L., Gross, M.L., Liu, Z., Faras, A.J., Hackett, P.B., Kapuscinski, A.R. and Guise, K.S. (1989) Transfer of the bovine growth hormone gene into northern pike, *Esox lucius*. J. Cell. Biochem. 13B: 173 (abstract).

Sekine, S., Mizukumi, T., Nishi, T., Kuwana, Y., Saito, A., Sato, M., Itoh, S. and Kawauchi, H. (1985) Cloning and expression of cDNA for salmon growth hormone in *Escherichia coli*, Proc. Natl. Acad. Sci., U.S.A. 82: 4306-4310.

Simons, J.P., Wilmut, I., Clark, A.J., Archibald, A.L., Bishop, J.O. and Lathe, R. (1988) Gene transfer into sheep. Bio/Technology 6: 179-183.

Small, J.A., Khoury, G., Jay, G., Howley, P.M. and Scangos, G.A. (1986) Early regions of JC virus and BK virus induce distinct and tissue-specific tumors in transgenic mice. Proc. Natl. Acad. Sci. USA 83: 8288-8292.

Song, S., Trinh, K. Y., Hew, C.L., Hwang, S.J., Belkhode,S. and Idler, D.R. (1988) Molecular cloning and expression of salmon prolactin cDNA. Eur. J. Biochem. 172: 279-285.

Soriano, P., Cone, R.D., Mulligan, R.C. and Jaenisch, R. (1986) Tissue-specific and ectopic expression of genes introduced into transgenic mice by retroviruses. Science 234: 1409-1413.

Southern, E.M. (1975) Detection of specific sequences among DNA fragments separated by gel electrophoresis, J. Mol. Biol. 98: 503-517.

Southern, P.J. and Berg, P. (1982) Transformation of mammalian cells to antibiotic resistance with a bacterial gene under control of SV40 early region promoter. J. Mol. Appl. Genet. 1: 327-341.

Stacy, N.E, Cook, A.F. and Peter, R.E. (1979) Spontaneous and gonadotrophin-induced ovulation in the goldfish, *Carassius auratus* L. Effects of external factors. J. Fish. Biol. 15: 349-361.

Steller, H., and Pirrotta, V. (1985) A transposable P vector that confers selectable G-418 resistance to *Drosophila* larvae. EMBO J. 4: 167-171.

Stuart, G.W., McMurray, J.V. and Westerfield, M. (1988) Replication, integration, and stable germ-line transmission of foreign sequences into early zebrafish embryos. Development 103: 403-412.

Svet-Moldavsky, G.J. (1958) Sarcoma in albino rats treated during the embryonic stage with Rous virus. Nature 182: 1452-1453.

Vielkind, J. R. (1992) Medaka and zebrafish: Ideal as transient and stable transgenic systems. In: Transgenic Fish. C. L. Hew and G. L. Fletcher (eds.) World Scientific Publishing Co., Singapore. pp. 72-91.

Watahiki, M., Tanaka, M., Masuda, N., Yamakawa, M., Yoneda, Y. and Nakashima, K. (1988) cDNA cloning and primary structure of yellowtail (*Seriola quinqueradiata*) pregrowth hormone. Gen. Comp. Endocrinol. 70: 401-406.

Weatherly, A.H. and Gill, J.A. (1987) Growth increases produced by bovine growth hormone in grass pickerel, *Esox americanus vermiculatus* (LeSueur), and the underlying dynamics of muscle fiber growth. Aquaculture 65: 55-65.

Wheaton, J.E., Al-Raheem, S.N., Massri, Y.G. and Marcek, J.M. (1986) Twenty-four hour growth hormone profiles in Angus steers. J. Anim. Sci. 62: 1267.

White, B.A. and Bancroft, F.C. (1982) Cytoplasmic dot hybridization. J. Biol. Chem. 257: 8569-8572.

Wilkie, T.M., Brinster, R.L. and Palmiter, R.D (1986) Germline and somatic mosaicism in transgenic mice. Develop. Biol. 118: 9-18.

Yoon, S.J., Hallerman, E.M., Gross, M., Liu, Z., Schneider, J.F., Faras, A.J., Kapuscinski, A.R. and Guise, K.S. (1990) Transfer of the gene for neomycin resistance into goldfish, *Carassius auratus*. Aquaculture 85: 21-33.

Zhang, P., Hayat, M., Joyce, C., Lin, C.M., Gonzalez-Villasenor, L.J., Dunham, R., Chen, T.T. and Powers, D.A. (1988) Gene transfer, expression, and inheritance of pRSV-trout-GH-cDNA in fish, First Intl. Symp. Marine Molec. Biotech. 9-11 Oct., 1988, Baltimore, MD (abstract).

Zhu, Z., Liu, G., He, L. and Chen, S. (1985) Novel gene transfer into the fertilized eggs of goldfish (*Carassius auratus* L. 1758). Z. Angew. Ichthyol. 1: 31-34.

Zhu, Z., Xu, K., Li, G., Xie, Y. and He, L. (1986) Biological effects of human growth hormone gene microinjected into the fertilized eggs of loach *Misgurnus anguillicaudatus* (Cantor). Kexue Tongbao 31: 988-990.

Zhu, Z. (1992) Generation of fast growing transgenic fish, methods and mechanisms. In: Transgenic fish. C. L. Hew and G. L. Fletcher (eds.) World Scientific Publishing Co., Singapore. pp. 92-119.

CHAPTER 9

Integration, Expression and Inheritance of Foreign Fish Growth Hormone Gene in Transgenic Fish

Thomas T. Chen, Chun-Mean Lin, Rex A. Dunham and Dennis A. Powers

ABSTRACT

Microinjection of DNA is currently employed as a standard technique to generate a range of transgenic animal species. This technique has also been adapted to produce transgenic fish with a high degree of success. Studies conducted by Agellon et al. (1988c) showed that administration of biosynthetic rainbow trout (rt) growth hormone (GH) polypeptide to juvenile rainbow trout resulted in a significant growth enhancement. These results point to the possibility of improving the growth rate of cultured fish via manipulation of growth hormone or its gene. As the first step toward reaching this goal, we have transferred the rtGH gene fused to the long terminal repeat (LTR) sequence of avian Rous sarcoma virus to common carp and channel catfish by microinjection. About 35% of the injected common carp embryos survived at hatching, and of which 10% of the survivors had stably integrated the rtGH gene. These transgenic individuals not only expressed the rtGH gene but also grew, on the average, 20% larger than their non-transgenic siblings. A significant fraction of the F_1 progeny derived from crosses between transgenic male and non-transgenic female carp inherited the rtGH gene. These transgenic progeny also grew faster than their non-transgenic siblings. These results suggest that, with further intensive research effort, it will be possible to produce fast growing fish strains for aquaculture by the gene transfer technology.

INTRODUCTION

As the world population increases rapidly, there is an urgent need to increase production of protein food sources. Fish products possess the greatest potential to meet this world demand. Traditionally, success in fisheries has been largely dependent upon the natural population of fresh-water and marine finfish, shellfish and macroalgae. However, due to increasing world consumption of fish products, as well as chemical pollution of the aquatic environments, and poor restocking effort and management, the total world-wide annual harvest level of fish products is rapidly approaching the maximal potential of between 100 to 150 million metric tons, as recently reported by the U.S. Department of Commerce and MOA. In 1986, the world-wide harvest of fish products reached 92 million metric tons. Recently a number of regions have shown a significant decline in the catches of important fish species such as salmon, striped bass, sturgeons, eels, jacks, mullets, mackerel, kris, abalone, oysters and crabs (Fisheries Dept., FAO Yearbook, 1988). As a result, fisheries have traveled farther to exploit more productive areas, switched to alternative species, and begun to employ sophisticated technology. This recent development has resulted in a significant increase in international fish prices. Since in recent years U.S. imports of fish products have risen almost twice as much as exports, such an increase in the international fish price has contributed significantly to the trade deficit. In fact, the U.S. trade deficit on fish products escalated to 6.6 billion U.S. dollars per year in 1988. In order to cope with these problems, dramatic measures for increasing fish production are required.

As a means of increasing the production of fish products, many countries have turned to aquaculture and mariculture. In 1985, the production of finfish, shellfish and macroalgae by aquaculture/mariculture reached 10.6 million metric tons, which was 12.3% of the world-wide catch generated by international efforts. Clearly aquaculture/mariculture has the potential to significantly increase the world production of fish products, and thus affect the trade balance of fish products.

Success in aquaculture and mariculture revolves around four key requirements: breeding, feeding, disease protection, and management. Applications of molecular biology and biotechnology can help to improve the means of satisfying the first three requirements. These applications include enhancing growth rates, controlling reproductive cycles, improving feeds, developing vaccines, developing disease resistance and hardier genetic stocks, and many other possibilities. Using economically important fish species as models, we are investigating strategies to increase growth rates of finfish and shellfish in aquaculture operations through manipulation of GH and growth factor genes. In this paper we will discuss results of our studies to increase growth rates of common carp and catfish by the gene transfer technology.

Growth Enhancement of Biosynthetic Rainbow Trout Growth Hormone

Through molecular cloning, two genes encoding growth hormone polypeptides (GH1 and GH2) were identified in rainbow trout, *Oncorhynchus mykiss* (Agellon and Chen, 1986; Agellon et al., 1988a,b). A large quantity of biologically active rainbow trout biosynthetic GH polypeptide was prepared by expressing rtGH1 cDNA sequence in E. coli cells (Agellon et al., 1988c). This biosynthetic hormone is a fusion protein, containing 9 amino acid residues derived from the N terminus of E. coli ß-galactosidase and the entire sequence of the mature GH polypeptide.

Agellon et al. (1988c) demonstrated in a series of bioassays that administration of this hormone to yearling rainbow trout resulted in a significant growth enhancement. After treatment of yearling rainbow trout with the biosynthetic GH for four weeks at a dose of 1 µg/g body weight/week, the weight gain in the individuals of the hormone-treated group was two times greater than that of their controls (Fig. 1). Significant length gain was also evident in hormone-treated animals. The muscle chemical composition of hormone-treated fish was not significantly different from that of the controls (Table 1). When the same biosynthetic hormone preparation was administered

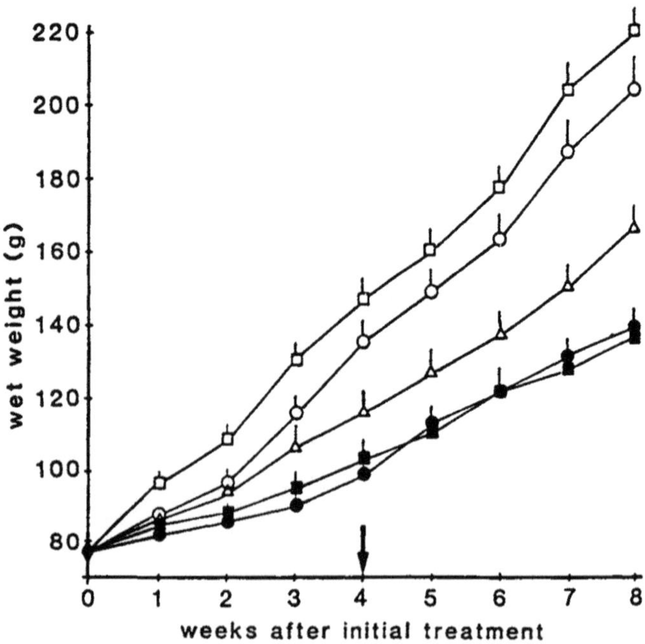

Fig. 1. Effect of biosynthetic trout GH on growth of yearling rainbow trout. Groups of yearling rainbow trout received intraperitoneal injections of biosynthetic GH or control extract for five weeks. Wet weights of GH-treated and control fish are shown (mean ± SE). Open symbols, GH-treated fish: △, 0.2 ug/g body weight; ○, 1.0 ug/g body weight; □, 2 ug/g body weight. Closed symbols, control fish: ●, mock-treated fish; ■, untreated fish. The arrow indicates the time of the last hormone treatment (Agellon et al., 1988c, with permission).

to rainbow trout fry by immersing the fish in a GH containing solution, the same growth-promoting effect was observed (Table 2; Agellon et al., 1988c; Leong and Chen, unpublished results). These results are in agreement with those reported by Sekine et al. (1985) and Gill et al. (1985) demonstrated that the application of biosynthetic GH from other sources to rainbow trout, chum salmon or coho salmon results in growth-enhancement. However, it is worth noting that the growth effect of the biosynthetic hormone was markedly reduced when more than 2 μg/g body weight of the hormone was applied to the test animals (Fig. 1). These results clearly suggest that

Table l. Percent of moisture, protein, lipid and ash in the muscle tissue of GH-treated and control fish (Agellon et al., 1988c with permission).

	Percent ± SE (n = 4)			
Treatment	Moisture	Protein	Lipid	Ash
Uninjected	78.58±1.4	16.54±1.0	3.58±0.5	1.31±0.1
Mock-treated	80.26±0.8	15.08±0.4	3.48±0.5	1.19±0.1
GH (0.2ug/g)	78.91±0.9	15.98±1.0	4.04±0.6	1.17±0.1
GH (l.Oug/g)	78.37±0.3	16.58±0.2	3.91±0.5	1.32±0.1
GH (2.Oug/g)	79.02±0.9	15.99±0.3	3.89±0.9	1.24±0.1

Yearling rainbow trout were treated weekly for 5 weeks with control and GH-containing bacterial extracts. Four weeks after the withdrawal of hormone treatment, tissue samples were collected from the same region of the carcass of sacrified fish. Chemical composition analysis was performed after tissue lyophilization. Values shown are corrected to the initial wet weight of the tissue sample.

Table 2. Effect of GH treatment on the growth of rainbow trout fry (Agellon et al., 1988c, with permission).

	Weight (gm ± SD)		
Treatment	Initial	Final	% Gain
Saline control	1.33±0.6**	3.94±1.8*	19
GH (50 ug/l)	1.29±0.7**	5.51±1.6***	327
GH (500 ug/l)	1.35±0.7**	5.30±1.3***	293

Groups of rainbow trout fry (n = 15) were subjected to osmotic shock in the presence or absence of GH. Weight was measured prior to and 5 weeks post-treatment. Differences between mean weights of GH-treated and control groups were evaluated using student's it test. Mean weights were considered to be significantly different if $p < 0.01$.

*Significantly different from the GH-treated groups ($p < 0.01$); **No significant difference between these groups; ***No significant difference between these two treatments.

when the total amount of GH exceeds the maximal threshold level, the homeostasis of the hormone will be disturbed, and consequently affect the performance of the animal.

Several years ago, Morse (1984) reported that bovine insulin and GH enhanced the growth of California red abalone. Recently Paynter and Chen (unpublished results) have observed that administration of biosynthetic rtGH polypeptide to spat of juvenile oysters (*Crassostrea virginica*) by the "dipping method" referred to above, also resulted in significant increases in shell height, shell length, shell weight, wet weight and dry weight.

Although results discussed above clearly point to the possibility of enhancing growth rates of cultured finfish and shellfish by supplying optimal levels of biosynthetic GH, many studies are still required to delineate a number of basic and practical scientific issues. These issues include: (i) efficient methods for large scale production, purification and renaturation of biosynthetic GH polypeptide; (ii) effective means of hormone delivery; (iii) detailed dose regimens for each target fish species; (iv) the effect of both chronic and acute GH treatment; (v) nutrient requirements; and (vi) genetic and physiological regulation of GH action.

GENERATION OF TRANSGENIC FISH

Although exogenous application of biosynthetic GH enhances growth rates of cultured fish, it may or may not be cost effective. The use of gene transfer technology to generate new strains of fish that express higher levels of GH, as an alternative, would bypass many of the problems associated with exogenous GH treatment. Moreover, once transgenic fish strains have been generated, they would be far more cost effective than their non-transgenic counterparts because these fish would have their own means of producing and delivering the hormone, and could transmit their enhanced growth characteristics to their offspring.

Zhu et al. (1985), in China, reported the first successful transfer of human GH gene fused to a mouse metallothionein gene promoter into goldfish. According to Zhu (personal communication)(See also Chapter 6 by Zhu), the F_1 offspring of these transgenic fish grew twice as large as their non-transgenic siblings. Although Zhu and his colleagues did not present compelling evidence for integration and expression of the foreign genes in their transgenic fish, many laboratories (including our groups) have successfully confirmed their work by demonstrating that the GH gene and other gene constructs can be transferred into embryos of a number fish species and integrated into the genomic DNA of the host fish (for review, see Chen and Powers, 1990). While a few groups have demonstrated expression of foreign genes in transgenic fish, we are the only group to document that foreign fish GH genes are: (a) transferred to the target fish species; (b) integrated into the fish genome; (c) genetically transmitted to the next (F_1) generation; and (d) the expression of the foreign fish GH gene increased growth rates of both P_1 and F_1 generations of transgenic fish (Zhang et al., 1990 and Chen et al., 1990).

In gene transfer studies conducted in our laboratories (Zhang et al., 1990 and Chen et al., 1990), about 10^6 molecules of a linearized recombinant plasmid containing the long terminal repeat (LTR) sequence of avian Rous sarcoma virus (RSV) and the rainbow trout GH1 cDNA were injected into the cytoplasm of one-cell, two-cell and four-cell common carp and catfish embryos. Genomic DNA samples extracted from the pectoral fin of individuals derived from these embryos were analyzed for the presence of RSVLTR-rtGHl-cDNA by dot blot and Southern blot hybridization, using radio-labelled LTR of RSV and/or trout GH1 cDNA as probes. Results of this study showed that about 35% of the injected embryos survived to hatching, of which about 10% of the survivors had stably integrated the pRSVLTR-rtGHl-cDNA sequence (Table 3). A similar percentage of transgenic fish was also obtained when the same GH gene construct was injected into catfish embryos (Powers et al., in press). Southern blot analysis of genomic DNA samples of several transgenic fish revealed that a single copy of the RSVLTR-rtGHl-cDNA sequence was integrated at multiple chromosomal sites (Fig.2). In some cases, the Hind III site residing between RSVLTR and rtGHlcDNA was modified. While foreign gene sequences introduced into embryos by micro-injection are usually integrated as a head-to-tail concatemer at a single chromosomal site and modified (Palmiter et al., 1982; Dunham et al., 1987), examples of single or low copy integrations of foreign gene sequences exhibiting deletions, modifications or rearrangements have also been observed (Gordon and Ruddle, 1985). Hence our results of pRSVLTR-rtGHl-cDNA integration in transgenic carp agree with those

Table 3. Percent of hatching, survival and integration of carp embryos microinjected with pRSVrtGHlcDNA at different developmental stages (Chen et al., 1990, with permission).

Embryo stages	Embryos injected	Hatched (%)	Survival at 90 days	Fingerlings analyzed	Integration (%)
One-cell	1034	39.3	52.7	217	9.9
Two-cell	331	33.3	70.6	771	5.6
Four-cell	381	33.0	58.4	73	1.4
Control (non-injected)	569	33.9			

reported in other systems.

Expression and Inheritance of Growth Hormone Gene in Transgenic Fish

Agellon et al. (1986a) reported earlier that polyclonal antibodies to chum salmon GH reacted specifically with the GH polypeptide of rainbow trout. Hence these antibodies had been used as probes for detecting the production of trout GH in transgenic carp by a quantitative dot blot immunobinding assay. Since these polyclonal antibodies showed a partial cross-reactivity with common carp GH, they were rendered specific to trout GH by extensive re-absorption with pituitary extracts of common carp. Trout GH was not detected in the circulation of P_1 transgenic carp, since the rtGH1-

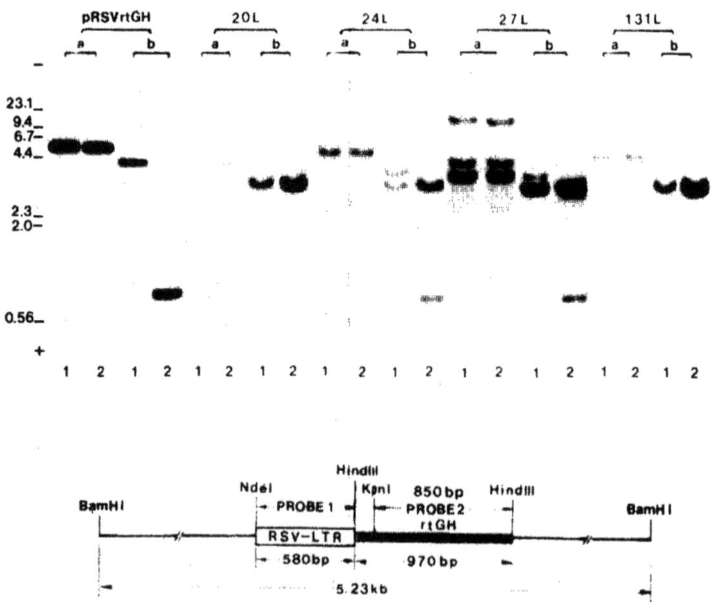

Fig. 2. Southern blot analysis of genomic DNA samples isolated from transgenic carp. (A). Genomic DNA samples isolated from pectoral fins of transgenic carp were digested with Bam-H1 (a) or Hind-III (b), electrophoresed, transferred to nylon membranes, and then hybridized to [32P]-labelled probe 1 (1) or probe 2 (2). Bam H1 and Hind-III digested pRSVrtGH1cDNA samples were used as positive controls. (B). A 580 bp Nde-I-Hind-III fragment derived from pRSVrtGH1cDNA was used as probe 1, and a 850 bp Kpn-I-Hind-III fragment derived from pRSVrtGH1cDNA was used as probe 2. (Chen et al., 1990, with permission)

cDNA used in the gene transfer studies does not contain a signal peptide sequence. Instead, various levels of rtGH were detected in the red blood cell (RBC) transfer of transgenic individuals: ranging from 8.0 to 89.1 ng/mg RBC proteins (11,12). As expected, there was no correlation between the number of pRSVLTR-rtGHl-cDNA integrated and levels of rtGH expressed in RBC (Table 4). These results are consistent with those in transgenic mice reported by Palmiter et al. (1982). Although there was considerable variation in size among the P_1 transgenic fish, they were 22%, on the average, larger ($p < 0.05$) than their sibling controls (Fig.3).

To study patterns of inheritance of RSVLTR-rtGHI-cDNA sequence in the transgenic carp sperm samples collected from several sexually mature P_1 male transgenic fish (04R, 36L, 131L and 94R) were crossed to eggs collected from one non-transgenic female, and DNA samples extracted from resulting F_1 progeny were assayed for the presence of RSVLTR-rtGHl-cDNA sequence by dot blot hybridization. None of the F_1 progeny derived from transgenic male 04R received the RSVLTR-rtGHl-cDNA sequence from their father, suggesting that the RSVLTR-rtGHl-cDNA sequence was not integrated into the germ line of fish 04R. Although most of the F_1 progeny derived from fish 36L died, the four survivors inherited the foreign gene sequence. Results obtained from F_1 progeny of fish 131L and 94R are most informative. About 31.3% and 42.3% of F_1 progeny derived from fish 131L and 94R respectively carried RSVLTR-rtGHl-cDNA sequence (Table 5). Since transgenic fish are independently derived by injecting the foreign gene at different stages of development, one would expect the transgenic animals to be mosaic for RSVLTR-rtGHl-cDNA. The degree of mosaicism in transgenic animals determines whether a foreign gene will be present in the germ line and whether it will be transmitted to subsequent generations. Since nearly 50% of the progeny derived from fish 94R carry RSVLTR-rtGHl-cDNA, the transformed progenitor cells must be primordial to the entire germ-line. Conversely, the progeny ratios of 1 transgenic : 3 non-transgenic from fish 131L suggests this germ-line is mosaic.

The body weights of F_1 progeny derived from fish 131L and 94R were measured at the age of three months. The positive F_1 progeny of fish 131L and 94R were 20.8% ($p < 0.05$) and 40. 1 % ($p < 0.00$ 1) larger than their non-transgenic siblings, respectively (Table 5). Furthermore, these progeny were 32% and 46% larger than their largest non-transgenic siblings respectively. These results clearly demonstrate that the growth enhancement phenotype resulting from the presence of the RSVLTR-rtGH1-

cDNA sequence in transgenic fish 131L and 94R has been passed down to their respective F_1 progeny through the germ-line.

Table 4. Gene copy number and rainbow trout growth hormone levels in the red blood cells of transgenic carp (Chen et al., 1990, with permission).

Fish number	pRSVrtGHcDNA molecules[a]	ng GH/mg protein in red blood cells
20L	2	8.0
34L	4	47.7
36L	1	47.5
94R	1	48.8
OR	3	64.2
04R	2	28.6
07R	1	8.9
08R	3	73.8
131L	2	81.9

[a]copy number per haploid genome.

Table 5. Mean weight, range weight and percent inheritance at 90 days of progeny derived from transgenic fish 131L and 94R (Chen et al., 1990, with permission).

	Progeny of fish 131L		Progeny or fish 94R	
	Transgenic	Non-transgenic	Transgenic	Non-transgenic
N	31	65	11	15
% inheritance	32.3		42.3	
Mean weight (g)	120.6[a]	99.3[b]	206	147
SE[c]	17.4	14.7	45.2	48
Weight range	95-173[d]	65-129	115-283[e]	67-228
% difference	20.8[f]		40.1[g]	

a: N = 28; b: N = 38; c: SE = standard error
d & e: 32%d and 46%e of transgenic progeny were larger than largest control.
f & g: Transgenic progeny were larger than non-transgenic progeny at $p < 0.05^f$ and $p < 0.001^g$ respectively.

Fig. 3. P_1 transgenic carp. The top two larger individuals are P_1 transgenic carp carrying and expressing the pRSVrtGH1cDNA while the bottom smaller individual is a P_1 nontransgenic sibling (Chen and Powers, 1990).

FUTURE PROSPECTS

Although results from our studies and those of others showed that transfer, expression and inheritance of a foreign gene in finfish has been achieved, considerable amounts of research are still required in order to develop strains of transgenic fish for commercial aquaculture/mariculture purposes. To achieve this ultimate goal of applying gene transfer technology in aquaculture/mariculture, several major breakthroughs are required. These include: (a) developing efficient mass gene transfer technologies; (b) developing rapid procedures for identifying presumptive transgenic individuals; (c) developing methods for targeting foreign gene integration into specific areas of the fish genome; (d) identifying suitable promoters to drive the expression of the foreign gene; (e) determining physiological, nutritional, immunological, and environmental factors that will maximize the performance of transgenic individuals; (f) assessing safety and environmental impacts of transgenic fish.

ACKNOWLEDGMENTS

This work is supported by grants from the Maryland Sea Grant Program (R/F 63), Office of Naval Research, and BARD to RAD, TTC and DAP.

REFERENCES

Agellon, L.B. and Chen, T.T. (1986). Rainbow trout growth hormone: molecular cloning and expression in E. coli. DNA 5: 463-471.

Agellon, L.B., Chen, T.T., van Beneden, R., Sonstegard, R.A., Wagner, G.F., and McKeown, B.A. (1986a). Rainbow trout growth hormone: in vitro translation of pituitary RNA and product analysis. Can. J. Fish. Aqua. Sci. 43: 1327-1331.

Agellon, L.B., Davies, S.L., Chen, T.T., and Powers, D.A. (1988a) Nucleotide sequence of a gene encoding growth hormone (GH): implication on the evolution of the GH gene structure. Proc. Natl. Acad. Sci. USA 85: 5136-5140 .

Agellon, L.B., Davies, S.L., Lin, C.M., Chen, T.T., and Powers, D.A. (1988b) Growth hormone in rainbow trout is encoded by two separate genes. Mol. Reproduc. Develop. 1: 11-17.

Agellon, L.B., Emery C.J., Jones, J.M., Davies, S.L., Dingle, A.D., and Chen, T.T. (1988c) Growth hormone enhancement by genetically engineered rainbow trout growth hormone. Can. J. Fish. Aquat. Sci. 45: 146-151.

Chen, T.T., Lin, C.M., Zhu, Z., Gonzalez-Villasenor, L.I., Dunham, R.A., and Powers, D.A. (1990) Gene transfer, expression and inheritance of rainbow trout and human growth hormone genes in carp and loach. In: Transgenic Models in Medicine and Agriculture. R. Church (ed.) Wiley-Liss, Inc., New York, pp. 127-139.

Chen, T.T., and Powers, D.A. (1990) Transgenic fish. Trends in Biotechnology 8: 209-218.

Dunham, R.A., Eash, J., Askins, J., Townes, T.M. (1987) Transfer of the metallothionein-human growth hormone fusion gene into channel catfish. Trans. Am. Fish. Soc. 116: 87-91.

Fisheries Department, FAO Yearbook, (1988) Food and Agriculture Organization of the United Nation, Vol. 62, 63.

Gill, J.A., Stumpter, J.P., Donaldson, E.M., and Dye, H.M. (1985) Recombinant chicken and bovine growth hormone in cultured juvenile Pacific salmon (*Onchorhynchus kisutch*). Bio/Technology 3: 4306-4310.

Gordon, J.W., and Ruddle, F.H. (1985) DNA-mediated genetic transformation of mouse embryos and bone marrow (a review). Gene 33: 121-136.

Morse, D.E. (1984) Biochemical and genetic engineering for improved production of abalone and other valuable molluscs. Aquaculture 39: 263-282.

Palmiter, R.D., Brinster, R.L., Manner, R.E., Trumbauer, M.E., Rosenfeld, M.G., Birnberg, N.C., and Evans, R.M. (1982) Dramatic growth of mice that develop from eggs microinjected with metallothionein-growth hormone fusion genes. Nature 300: 611-615.

Powers, D.A., Gonzalez-Villasenor, L.I., Zhang, P., Chen, T.T., and Dunham, R.A. (1991) Studies on Transgenic fish: gene transfer, expression and inheritance. In Transgenic Animals. N. First and F. Haseltine (eds), Butterworth-Heinemann, 307-323.

Sekine, S., Miizukzmi, T., Nishi, T., Kuwana, Y., Saito, A., Sato, M., Itoh, H., and Kawauchi, H. (1985) Cloning and expression of CDNA for salmon growth hormone in *E. coli*. Proc. Natl. Acad. Sci. USA 82: 4306-4310.

Zhu, Z., Li, G., He, L., and Chen, S. Z. (1985) Novel gene transfer into the goldfish (*Carassius auratus* L. 1758). Angew Ichthyol. 1: 31-34.

Zhang, P., Hayat, M., Joyce, C., Gonzalez-Villasenor, L.I., Lin, C.M., Dunham, R.A., Chen, T.T., and Powers, D.A. (1990) Gene transfer, expression and inheritance of pRSV-rainbow trout-GHcDNA in the common carp, *Cyptinus carpio*.Molec. Reproduc. Develop. 25: 13-25.

CHAPTER 10

Growth Hormone Gene Transfer in Atlantic Salmon: Use of Fish Antifreeze/Growth Hormone Chimeric Gene Construct

Shao Jun Du, Zhiyuan Gong, Garth L. Fletcher, Margaret A. Shears and Choy L. Hew

ABSTRACT

The promoter activity of the antifreeze polypeptide (AFP) gene from the ocean pout was assayed in three salmonid cell lines. The 5' upstream sequence (2.2 kb) from a genomic AFP clone was fused to a bacterial chloramphenicol acetyl transferase (CAT) gene. After transfection, CAT activities were demonstrated in all three cell lines tested. These experiments suggest that the AFP promoter is active and suitable for gene transfer in salmonids. The AFP promoter was fused to a growth hormone cDNA clone from the chinook salmon. Linearized AFP-GH gene was microinjected into the fertilized, nonactivated Atlantic salmon eggs. The presence of transgenics was analyzed by polymerase chain reactions using specific primers. Out of 450 fingerlings analyzed, nine positives were identified.

INTRODUCTION

The rate at which fish grow from egg to market size is an important component to the economic viability of any commercial aquaculture venture. Therefore, procedures that can accelerate growth could be of considerable value to the aquaculture industry. Despite the body of research demonstrating the growth-promoting effects of parenterally administered mammalian, avian, piscine and fish growth hormones (GH) on fish (Higgs et al. 1976; Gill et al. 1985; Sekine et al. 1985; Agellon et al. 1988), as of yet there is no viable method of delivering the hormone to the fish under commercial culture conditions.

One solution to the problem of administering growth hormone to fish lies with gene transfer. The potential of this procedure was illustrated dramatically by Palmiter et al. (1982) when they demonstrated that mice could be rendered giants by the addition of a few extra growth hormone genes to their genome.

One of the problems associated with growth hormone gene transfer is the necessity of modifying its tissue specific expression. There is little value in transferring growth hormone genes if they can only be expressed in the pituitary gland under homeostatic control via the hypothalamus (Lira et al. 1988). In order to solve this problem investigators construct chimeric growth hormone genes using promoters that would be expected to be expressed in tissues other than the pituitary gland. Up until the present time the promoters that have been used include mouse metallothionein-1 (mMT-1), Rous sarcoma virus (RSV) and SV40.

Since the main reason for producing transgenic fish is to improve the economics of fish production for food it seems unwise to use gene constructs from organisms other than fish (human, bovine, mice, bacteria, viruses, etc.) because they are likely to be rejected at the market place. For this reason we have isolated the promoter region from one of our fish antifreeze protein genes and linked it to a full-sized chinook salmon cDNA clone.

This study reports on the analysis of the ocean pout (*Macrozoarces americanus*) antifreeze (AFP) gene promoter and the successful production of transgenic Atlantic salmon using an ocean pout AFP promoter chinook salmon cDNA gene construct. PCR protocols were developed for the routine analysis of large numbers of small blood samples.

MATERIALS AND METHODS

Experimental Materials

Mature Atlantic salmon *(Salmo salar)* were captured 2-3 weeks prior to spawning from the Exploits and Colinet river systems, Newfoundland, and transported live to the Ocean Sciences Centre, Memorial University of Newfoundland. The fish were maintained at seasonally ambient photoperiod in 2 x 2 x 0.5 m aquaria supplied with flowing freshwater and air.

Eggs and sperm were stripped from salmon that had been anaesthetized in a dilute solution of t-amyl alcohol. Eggs were fertilized up to 2h prior to microinjection and rinsed with several changes of ice-cold salmon Ringer solution. When this procedure is used the eggs are fertilized but not activated. Activation occurs when the eggs are placed in a hypotonic solution (such as water) after microinjection.

Plasmid Construction

The construction of AFP-GH chimeric gene was described in detail by Du et al. (1992). Briefly, a 2.2 kb Bam H1-Bgl II fragment containing the promoter sequence from the ocean pout AFP gene (clone λOP5, Hew et al. 1988) was ligated to a synthetic

73 bp Bgl II-Pst I fragment that contains the chinook salmon GH gene 5'-untranslated sequence. The ligated Bam H1-Pst I fragment was then ligated with 709 bp Pst I-Stu I fragment of the chinook salmon GH cDNA clone which contains the entire GH coding sequence and part of the 3'-untranslated region (Hew et al. 1989). The resultant plasmid was cut with Stu I and Aat II, and then ligated to a Hpa I-Hind III fragment from λOP-5 plasmid to include the ocean pout AFP gene polyadenylation and the transcription termination signals (Hew et al. 1988). The final construct was referred to as opAFP-GHc (Fig. 1). The same 2.2 kb Bam H1-Bgl II fragment containing the ocean pout AFP promoter sequence, as used for opAFP-GHc, was subcloned into pBLCAT 3 (Luckow and Schutz 1987). The resulting plasmid was referred to as op5B. Supercoiled plasmids for transfection were prepared by ethidium bromide CsCl gradient centrifugation.

Transient CAT Expression Assay

RTH-149, a rainbow trout hepatoma cell line and CHSE-214, a chinook salmon embryonic cell line, were kindly provided by Dr. L. Gedamu, University of Calgary, Canada. CHH-1, a chum salmon heart cell line was purchased from ATCC. The cells were maintained at 18°C in minimum essential medium supplemented with 25 mM HEPES buffer (Gibco). The fish cells were transfected by the calcium phosphate co-precipitation method with glycerol shock and CAT assay were carried out

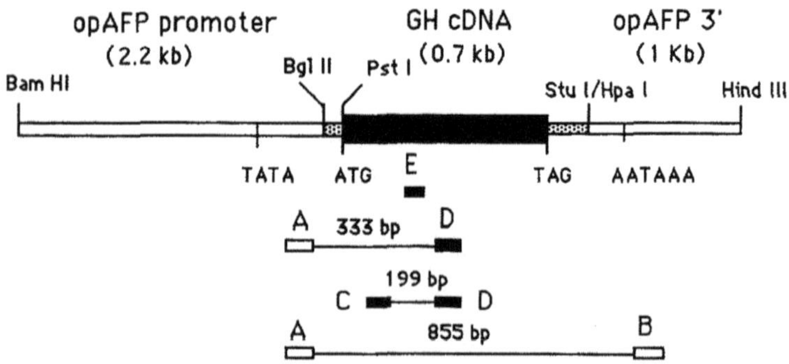

Fig. 1 Construction of the opAFP-GHc chimeric gene for transgenic studies. The GH cDNA sequence is indicated in box (dark, coding region; grid, untranslated region) and the DNA sequences from the ocean pout AFP gene are indicated by the open box. The TATA box, initiation site, stop codon and polyadenylation signal are indicated. The positions of PCR primers A, B, C and D, GH-probe E and the length of PCR products are also shown.

essentially according to Davies et al. (1986) and as modified for fish cells by Zafarullah et al. (1988). The detailed methods for cell culture and transient CAT expression assay were described by Gong and Hew (1991a).

Gene Transfer by Microinjection

The 4kb insert in opAFP-GHc was excised by EcoR I digestion and dissolved in saline buffer (0.9% NaCl) at a concentration of 5 μg/ml. Approximately 10^6 copies (2-3 nl) of the DNA insert was injected into a fertilized, but nonactivated salmon egg. The precedure is described in detail by Shears et al. (1991,Chapter 3,this monograph).

Synthesis of Deoxyoligonucleotide Primers for PCR

For PCR analysis, four primers were synthesized by the Biotechnology Service Centre, Hospital for Sick Children, Toronto, Canada. As shown in Figure 2, primer A, located at position +27 to +47 relative to the TATA box, is from the sense strand of the opAFP gene promoter; Primer B, located at position +861 to +881 relative to the TATA box, is from the antisense strand of the opAFP gene 3' flanking region. Primer C, located at position +161 to +181, is from the sense strand of GH coding sequence; While primer D, located at position +339 to + 359, is from the antisense strand of GH gene.

Primer A	5'	-GTCAGAAGTCTCAGCTACAGC-	3'
Primer B	5'	-ATCTCAACAGTCTCCACAGGT-	3'
Primer C	5'	-TCTGCTGATGCCAGTCTTACT-	3'
Primer D	5'	-ACAGAAGTCCAGCAGGAATAT-	3'

PCR Screening of Blood Samples

Thirty microliter blood samples were collected from 450 one year old fry. Three microliter of the blood was lysed in 100μl of 10mM NaOH, boiled in a water bath for 3 min., then centrifuged for 3 min. Three microliters of the supernatant were used for PCR analysis. PCR was carried out in 100 μl of reaction solution containing 50 mM KCl, 10 mM Tris, 1.5 mM $MgCl_2$, 0.1% Triton X-100, 1 μM of each primer, four deoxyribonucleotide triphosphate at 200 μM each, and 2.5 units of Taq DNA polymerase (Promega), 100 μl of mineral oil was added to prevent condensation. Amplification was started by denaturing at 92°C for 3 min. and followed by 30 cycles. Each cycle included 1 min. at 92°C (denaturation), 1 min. at 60°C (annealing), and 2 min. at 72°C (extension). After the final cycle, the temperature was held at 72°C for 10 minutes in order to complete the reaction. PCR was carried out by using the PTC-100 Programmable Thermal Controller (MJ Research Inc., Ocala, FL).

Analysis of the PCR Products by Agarose Gel and Southern Blot

Twenty microliters of the amplified product was subjected to electrophoresis on a 0.8% agarose gel, and the DNA products were visualized by ethidium bromide staining. The DNA was transferred to a Nylon membrane (Amersham) for Southern analysis. A 17 nt GH-specific probe E, 5' -GAAAATGTTCAATGACT- 3', from sequence of chinook salmon GH cDNA sense strand (+278 to +294 to the TATA box in the opAFP-GHc transgene) (Fig. 1) was end labelled with ^{32}p by T4 kinase and used as a probe for hybridization.

RESULTS

1. Ocean pout AFP promoter can function in salmonid cells *in vitro*

To test the effectiveness of the ocean pout AFP promoter, the Bam H1-Bgl II fragment as described in Materials and Methods, was subcloned into a CAT vector. The resulting construct, op5B, was tested for CAT activity in cultured fish cells. Three salmonid cell lines were used in this study: RTH-149, rainbow trout hepatoma; CHSE-214, chinook salmon embryonic cells; and CHH-1, chum salmon heart cells. As shown in Fig. 2, when op5B was transfected into the three cell lines, CAT activity was clearly detected in all of them. The level of CAT activity resulting from op5B is comparable to that from pBLCAT2, which has the strong thymidine kinase promoter from Herpes simplex virus. However, when these cells were transfected with pBLCAT3, a promoterless CAT construct, little or no CAT activity was detected. Similarly, the ocean pout AFP construct, op5B was active in Japanese medaka embryos (Gong et al. 1991b). These results indicate that the promoter from ocean pout AFP is functional in both cell lines and fish embryos, thus confirming its usefulness for transgenic studies.

2. The PCR-based screening strategy

Some of the problems associated with the use of PCR to screen for the presence of transgenes include observations of false positive and false negative signals. This necessitates careful selection of primers that will not only correctly identify the presence of a transgene but will also identify the presence of a preselected endogenous gene.

In the present study three sets of primers were used to screen for the presence of the chimeric AFP-GH gene in Atlantic salmon developed from microinjected eggs: primers A/B, A/D and C/D. Primers A and B are specific for the 5' and 3' ends respectively of the AFP gene, thus, they are absent in wild-type Atlantic salmon. The PCR technique would be expected to generate an 855 bp DNA fragment only in transgenic animals.

Primer D is derived from the chinook salmon GH cDNA gene, thus it would be expected to hybridize with the endogenous Atlantic salmon GH gene. However, when primer D is used in combination with primer A, which is specific for the 5' end of the AFP gene, only DNA from salmon transgenic for opAFP-GHc gene will be amplified to generate a 333 bp fragment.

Primers C and D are both derived from the chinook salmon GH cDNA gene. Thus they will hybridize to the endogenous Atlantic salmon GH gene and to the exogenous chinook salmon GH DNA when present. Since there is an intron between primers C and D in the salmon GH gene, a 344 bp fragment will be generated from all salmon DNA samples. However, if the opAFP-GHc is present as transgene, a shorter 199 bp fragment will also be generated. The C/D primers provide us with a positive control for the PCR analysis while at the same time screening for the presence of the exogenous opAFP-GHc insert.

3. The identification of transgenic salmon by PCR

Eight month old Atlantic salmon fry developed from the eggs injected in November 1989 with opAFP-GHc were bled in October, 1990. The DNA from the nucleated red blood cells were used directly for PCR analysis using primer A/D to determine the presence of the opAFP-GHc transgene. Out of the 450 fish analyzed, nine were shown to be positive. The results for six of them are illustrated in Figure 3. A 333 bp fragment was generated from salmon #14, #20, #28, #31, #34, #42 and the positive control (opAFP-GHc plasmid) (Fig. 3A). In contrast, this 333 bp fragment was absent in the noninjected salmon. The size of the amplified fragment (333 bp) corresponded with this size predicted from the transgene sequence.

To confirm that the amplified 333 bp DNA fragment was derived from the transgene opAFP-GHc, the DNA was transferred to a Nylon membrane for Southern blot analysis and hybridized with a GH specific probe E (Figure 1) derived from part of the sequence between primers C and D (Fig. 3B). As expected, the PCR fragment hybridized with the GH probe.

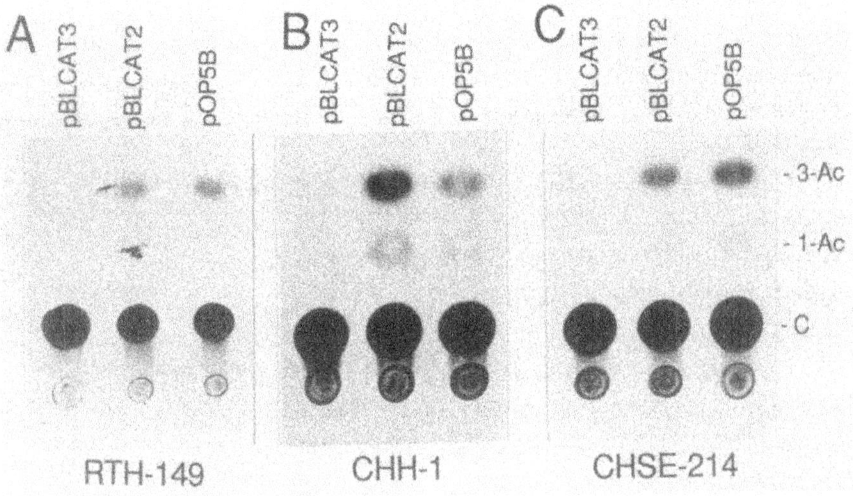

Fig. 2. CAT assays for ocean pout AFP promoter in cultured fish cells. RTH-149 (A), CHH-1 (B) and CHSE-214 (C) were transiently transfected with op5B (ocean pout AFP promoter), pBLCAT2 (viral thymidine kinase promoter) and pBLCAT3 (promoterless control), respectively. Acetylated and non-acetylated chloramphenicols, separated by thin layer chromatography, are indicated on the right.

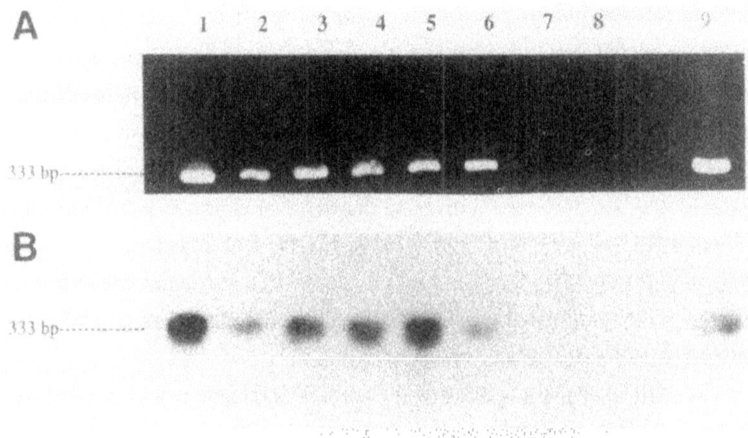

Fig. 3 (A). Identification of transgenic fish by PCR using primers A/D. Lane 1-6, transgenic fish #14, #20, #28, #31, #34 and #42; lane 7, H_2O; lane 8, control Atlantic salmon DNA; lane 9, opAFP-GHc plasmid. (B). Autoradiographs of the PCR product with GH-specific probe E.

To further confirm the presence of transgene opAFP-GHc, the DNA samples from #31 and #34 were amplified using primers C/D. As discussed in the screening strategy, the expected 344 bp DNA fragment derived from the endogenous Atlantic salmon GH gene was found in all of the salmon including the non-injected control analyzed. In addition, a smaller DNA fragment (199 bp) was found only in the transgenic fish. This 199 bp fragment was derived from the chinook salmon GH cDNA gene in the transgene. The identity of the bands was confirmed by hybridizing with GH-probe E. These results are shown in Figure 4.

4. The GH coding regions of the transgenes are intact in the transgenic salmon

To test the integrity of the GH coding sequence of the transgene in the transgenic

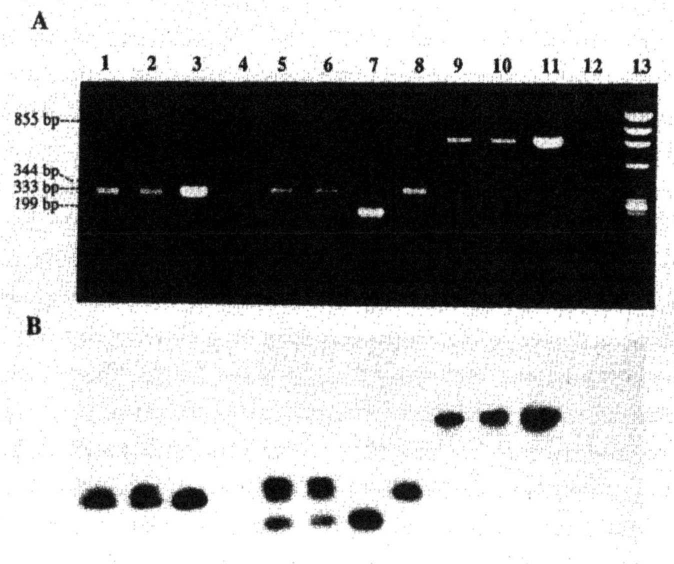

Fig. 4. (A). Agarose gel electrophoresis of PCR amplification products using 3 different sets of primers. Primers A/D, lanes 1-4; primers C/D, lanes 5-8; primers A/B, lanes 9-12. Lanes 1, 5 and 9, transgenic fish #31; lanes 2, 6 and 10, transgenic fish #34; lane 3, 7 and 11, opAFP-GHc plasmid; lane 4, 8 and 12, control Atlantic salmon DNA; lane 13 DNA molecular weight marker (ϕX-174-RF DNA Hae III digest). (B) Southern blot using GH specific probe E.

fish, primers A/B were used for PCR analysis. An 855 bp fragment was found in all nine positive fish. The fragment hybridized with the GH-probe E, indicating that the GH cDNA gene sequence in the transgene was intact in the transgenic salmon. The results, using #31 and #34 as examples, are shown in Figure 4.

DISCUSSION

We have been studying the physiology, biochemistry and molecular biology of fish antifreeze proteins for the past 15 years (Davies et al. 1988, 1989, Davies and Hew 1990). Therefore, it seemed appropriate to isolate the promoter regions of the antifreeze genes with a view to using one or more of them in the construction of chimeric gene constructs for aquaculture.

Antifreeze proteins have been isolated from the blood plasma of a number of cold water marine teleosts. They are synthesized predominantly in the liver and secreted into the blood where they lower the freezing temperature of the fish to values approaching the freezing temperature of seawater (-1.8° C). Their principal function is to protect the fish from freezing during the winter. Based on their structure, these proteins can be classified into antifreeze glycoproteins (AFGP) and antifreeze proteins (AFP). The AFP can be further divided into three distinct types. The type I, alanine-rich a helical AFP from winter flounder *(Pseudopleuronectes americanus)* and shorthorn sculpin *(Myoxocephalus scorpius)*, type II cystine-rich β-structured AFP from the sea raven *(Hemitripterus americanus)* and the type III AFP from ocean pout *(Macrozoarces americanus)* which has no distinguishing structural feature. The gene structures of these AFP and the activities of their promoters have been characterized in a series of manuscripts (Davies et al. 1989, Gong et al. 1991a and 1991b). The promoter regions of all of them are potentially valuable in the construction of chimeric genes. In addition, we have successfully transfered the AFP gene from the winter flounder into the Atlantic salmon by gene microinjection (see Fletcher et al., 1992, Chapter 11, this monograph).

We chose the promoter region of the ocean pout antifreeze gene to drive the chinook salmon GH cDNA gene for a number of reasons. In the first instance, the ocean pout antifreeze genes are expressed predominantly in the liver. Secondly, these genes are constitutively expressed throughout the year (Fletcher et al. 1985). This contrasts to the winter flounder antifreeze genes which are only expressed during the winter and are repressed by circulating blood growth hormone levels during summer (Fletcher et al. 1989). Another important feature of the ocean pout promoter sequence is that it is not present within the salmon genome. This greatly simplifies the PCR or genomic Southern blotting procedures used for screening salmon for the presence of the transgenes.

The results of the present study clearly indicate that the ocean pout antifreeze gene promoter region is capable of driving the expression of CAT reporter genes in cultured salmon cell lines. This result has been confirmed by Gong et al. (1991b) who demonstrated expression of the same chimeric construct in microinjected Japanese medaka *(Oryzias latipes)* embryos. Thus, there is no doubt that the ocean pout AFP promoter is functional and capable of driving the expression of growth hormone genes in transgenic fish.

Up until recently, the preferred method used for screening individual fish for the presence of transgenes was genomic Southern blotting. However, the exacting nature of this procedure rendered it too time consuming and tedious to be used routinely. Thus, screening a thousand individuals developed from microinjected eggs was all but impossible. The advent of PCR technology has solved most of the problems associated with identifying transgenic organisms. All that is required is a minimal amount of DNA (~0.01 μg), and this can be readily obtained from a small sample of whole blood (Chen and Evans, 1990).

Despite its simplicity, PCR is subject to a number of technical problems that can result in a failure to detect the presence of transgenes (false negatives), or the false identification of transgenic individuals (false positive). False positives are obtained by contamination of the buffers etc. by one or a few molecules of the gene construct under study. False negatives result from a failure of the PCR to amplify the sequence of interest.

In order to reduce the potential for technical problems, it is important that the primers be as specific as possible. In general, primers should be 20 to 30 nucleotides in length and have no more than 50-60% G/C content. The purines and pyrimidines should be randomly distributed to avoid stretches of G or C and significant secondary structure. Several programs are now available to aid in the selection of primer sequences (Lowe et al., 1990; Primer Designer 1990).

The quality of the DNA also affects PCR results. Although DNA isolated using phenol/chloroform extraction procedures is quite suitable, it is inappropriate for very small sample sizes. Recently Chen and Evans (1990) reported that small samples of blood cells from mice can be lysed by SDS and proteinase K and used directly for PCR. Another method developed by Davies et al. (1992, Chapter 4, this monograph) demonstrated that cells can be lysed by the PCR buffer or 10 mM NaOH without using SDS. Our own results demonstrate that DNA isolated from red blood cells using NaOH can be used directly for PCR analysis.

Considerable caution should be exercised to avoid contamination of the PCR reaction with the DNA sequences that are being screened for. Where possible, the PCR procedures should be carried out in locations free of contamination. In addition, pipettes, pipette tips, reagents and buffers used in association with PCR procedures should be used solely for this purpose. Clearly, once the transgene sequence has been amplified it is a potential source of contamination throughout the laboratory. An excellent technical report dealing with procedures to avoid false positive PCR results has been discussed by Kwok and Higuchi (1989).

The best method of avoiding false negatives is to include in the reaction mixtures of primers for a gene that is normally present in the organism. In our procedures we took advantage of the absence of antifreeze gene promoters in salmon and the difference in size between the exogenous GH cDNA sequence and the endogenous GH genomic DNA sequence.

Although the PCR technique is an invaluable tool to screen for transgenes it should be emphasized that it cannot determine whether the exogenous gene is integrated into a chromosomal site or merely exists within the cells as extrachromosomal DNA. Genomic Southern techniques are still the method of choice if evidence of genomic integration is desired.

The present study indicates that 2% (9 out of 450) of the salmon eggs injected with the opAFP-GHc gene retained the sequence following one year of development. This putative integration rate is essentially the same as we have observed for our winter flounder antifreeze gene construct (Shears et al. 1991a).

We are continuing to screen the remainder of our AFP-GH gene injected salmon (~2000 fish). Once this has been done we will be able to determine whether the presence of this exogenous gene is correlated with increased growth rates.

Note added in proof: Since the submission of this communication, we have completed the growth performance of the transgenic salmon. A number of these fish shows enhanced growth, with a 4 to 6 fold increase (Du et al., 1992).

ACKNOWLEDGMENTS

We wish to thank Madonna J. King for assistance in microinjection, Valda Boland for culturing the salmon, and the Biotechnology Service Centre (The Hospital for Sick Children, Toronto) for the synthesis of primers. This research was supported by NSERC strategic grant (to C.L.H. and G.L.F.) and a DFO studentship (to S.J.D.).

REFERENCES

Agellon, L. B. Emery, C. J., Jones, J. M., Davies, S. L., Dingle, A. D. and Chen, T. T. (1988) Promotion of rapid growth of rainbow trout (*Salmo gairdneri*) by a recombinant fish growth hormone. Can. J. Fish Aquat. Sci. 45: 146-151.

Chen, S. and Evans, G. A. (1990) A simple screening method for transgenic mice using the polymerase chain reaction. BioTechniques 8: 32-33.

Davies, L. G., Dibuer, M. D. and Battey, J. F. (1986) Basic Methods in Molecular Biology. Elsevier Science Publishing Co., pp. 298-300.

Davies, P. L., Hew, C. L. and Fletcher, G. L. (1988) Fish antifreeze proteins: physiology and evolutionary biology. Can. J. Zool. 66: 2611-2617.

Davies, P. L., Fletcher, G. L. and Hew, C. L. (1989) Fish antifreeze protein genes and their use in transgenic studies. In: Oxford Surveys on Eukaryotic Genes (Maclean, N., ed.) 6, 85-109: Oxford University Press.

Davies, P. L. and Hew, C. L. (1990) Biochemistry of fish antifreeze proteins. FASEB J. 4: 2640-2648.

Davies, P. L. and Gauthier, S. (1992) The application of PCR to aquaculture. In: Transgenic fish. C. L. Hew and G. L. Fletcher (eds) World Scientific Publishing Co, Singapore. pp. 61-71.

Du, S. J., Gong, Z., Fletcher, G. L., Shears, M. A., King, M. J., Idler, D. R. and Hew, C. L. (1992) Growth enhancement in transgenic Atlantic salmon: Use of all fish chimeric gene construct. Bio/Technology, In press.

Fletcher, G. L., Hew, C. L., Li, X., Haya, K. and Kao, M. H. (1985) Year round presence of high levels of plasma antifreeze peptides in a temperate fish: Ocean pout, *Macrozoarces americanus*. Can. J. Zool. 63: 488-493.

Fletcher, G. L., Idler, D. R., Vaisius, A. and Hew, C. L. (1989) Hormonal regulation of antifreeze protein gene expression in winter flounder. Fish Physiol. and Biochem. 7: 387-393.

Gill, J. A., Sumpter, J. P., Donaldson, E. M., Dye, H. M., Souza, L., Berg, T., Wypych, J. and Langley, K. (1985) Recombinant chicken and bovine growth hormones accelerate growth in aquacultured juvenile Pacific salmon, *Oncorhynchus kisutch*. Bio/Technology 3: 643-646.

Gong, Z. and Hew C.L. (1991a) Functional analysis of fish antifreeze gene promoters: Presence of negative regulatory regions. Submitted.

Gong, Z., Vielkind, J. R. and Hew, C. L. (1991b) Functional analysis and temporal expression of promoter regions from fish antifreeze protein genes in transgenic Japanese Medaka embryos. Mol. Marine Biol. Biotech. 1: 64-72.

Hew, C. L., Wang, N-C., Joshi, S., Fletcher, G. L., Scott, G. K., Hayes, P. H., Buettner, B. and Davies P. L. (1988) Multiple genes provide the basis for antifreeze protein diversity and dosage in the ocean pout, *Macrozoarces americanus*. J. Biol. Chem. 263: 12049-12055.

Hew, C. L., Trinh, K. Y., Du, S. J. and Song, S. (1989) Molecular cloning and expression of salmon pituitary hormones. Fish Physiol. Biochem. 7: 375-380.

Higgs, D. A., Donaldson, E. M., Dye, H. M. and McBride, J. R. (1976) Influence of bovine growth hormone and L-thyroxine on growth, muscle composition and histological structure of the gonads, thyroid, pancreas, and pituitary of coho salmon *(Oncorhynchus kisutch)*. J. Fish. Res. Board Can. 33: 1585-1603.

Kwok, S. and Higuchi, R. (1989) Avoiding false positives with PCR. Nature 339: 237-238.

Lira, S. A., Crenshaw III, E. B., Glass, C. K., Swanson, L. W. and Rosenfeld, M. G. (1988) Identification of rat growth hormone genomic sequences targeting pituitary expression in transgenic mice. Proc. Natl. Acad. Sci. USA 85: 4755-4759.

Lowe, T., Sharefkin, J., Yang, S. Q. and Dieffenbach, C. W. (1990) A computer program for selection of oligonucleotide primers for polymerase chain reactions. Nucleic Acids Res. 18: 1757-1761.

Luckow, B. and Schutz, G. (1987) CAT constructions with multiple unique restriction sites for the functional analysis of eucaryotic promoters and regulatory elements. Nucleic Acids Res. 15: 5490.

Palmiter, R. D., Brinster, R. L., Hammer, R, E., Trumbauer, M. E., Rosenfeld, M. G., Birnberg, N. C. and Evans, R. M. (1982) Dramatic growth of mice that develop from eggs microinjected with metallothionein-growth hormone fusion genes. Nature 300: 611-615.

Primer Designer (1990) PCR and oligo primer design program. Scientific & Educatioal Software. P. O. Box 440, State Line, PA, 17263, USA.

Sekine, S., Mizukami, T., Nishi, T., Kuwana, Y., Saito, A., Sato, M., Itoh, S., Kawauchi, H. (1985) Cloning and expression of cDNA for salmon growth hormone in *Escherichia coli*. Proc. Natl. Acad. Sci. U.S.A. 82: 4306-4310.

Shears, M. A., Fletcher, G. L., Hew, C. L., Gauthier, S. and Davies, P. L. (1991a) Transfer, expression, and stable inheritance of antifreeze protein genes in Atlantic salmon *(Salmo salar)*. Mol. Marine Biol. Biotech. 1: 58-63.

Shears, M. A., King, M., Goddard, S. V. and Fletcher, G. L. (1991) Gene transfer in salmonids by injection through the micropyle. In: Transgenic fish. C. L. Hew and G. L. Fletcher (eds.) World Scientific Publishing Co, Singapore. pp. 44-60.

Zafarullah, M., Bonham, K. and Gedamu, L. (1988) Structure of the rainbow trout metallothionein B gene and characterization of its metal-responsive region. Mol. Cell. Biol. 8: 4469-4476.

CHAPTER 11

Genetic Engineering of Freeze-Resistant Atlantic Salmon

Garth L. Fletcher, Peter L. Davies and Choy L. Hew

ABSTRACT

Winter flounder are protected from freezing in ice-laden seawater by high concentrations of antifreeze proteins (AFP) that circulate in the blood. Atlantic salmon do not have the benefit of protection from these macromolecular antifreezes and will freeze at temperatures below -0.7°C. The lack of AFP in salmon poses a severe limitation to their aquaculture in sea-pens along the Atlantic coast. In an effort to improve the freezing resistance of this species we have transferred an AFP gene from winter flounder to salmon. Linearized AFP gene was microinjected through the micropyle of fertilized but non-activated salmon eggs. When the blood from two year-old P_1 fish were assayed for the AFP gene by polymerase chain reaction (PCR) approximately 3% were found to be postive for the presence of the transgene. 40% of the transgenic salmon expressed the antifreeze protein precursor and secreted it into the blood plasma. Inheritance and expression of the AFP gene in the F_1 generation were established by PCR assays and immunoblotting. Crosses between two transgenic males and wild-type females yielded 15% and 64%, respectively, transgenic offspring (F_1). These results are consistent with the P_1 parents being germ-line mosaics for the incorporation of one or several copies of the flounder AFP gene. All of the F_1 transgenic salmon examined, expressed low levels of AFP in their blood. Approximately 50% of the offspring derived from four separate crosses between transgenic F_1 males and wild-type females were positive for the AFP gene. These results indicate the establishment of stable transgenic lines of Atlantic salmon.

INTRODUCTION

The freezing point of most aqueous solutions depends upon the concentration of dissolved solute particles (colligative properties). The higher the concentration of solute particles the lower the freezing point. Thus distilled water freezes at 0°C, whereas the body fluids of most fishes freeze at -0.5 to -0.8°C depending on their salt content. This means that fishes inhabiting fresh water are in no danger of freezing, unless, of course the water turns solid. Seawater has a salt concentration that is approximately three times that of fish body fluids. Therefore, its freezing temperature ranges from -1.7 to -2°C, depending on salinity.

Since the seawater temperatures in the polar oceans and along the Northeast Atlantic coast of Canada approach freezing for at least part of the year, why do not fishes inhabiting these waters freeze and die? Scholander and his colleagues first attempted to answer this question by studying a number of marine fish inhabiting the coast of Northern Labrador (Scholander et al., 1957; Gordon et al., 1962). They made two important observations. Some fish do survive supercooled; that is, at temperatures below their freezing point. If they come into contact with ice, however, they immediately freeze and die. Therefore, it is the combination of supercooling and ice contact that is lethal. The second observation made by Scholander's group was that the body fluids of some fish had the same freezing point as seawater. They termed the responsible solute "antifreeze".

The early work of DeVries and Feeney (DeVries et al., 1970) established that the antifreezes were polypeptides. Since that time, four distinct classes of antifreeze polypeptides differing in carbohydrate content, amino acid composition, protein sequence and secondary structure have been isolated from the blood sera of a diverse group of marine teleosts. These proteins are synthesized predominantly in the liver for export to the blood, which distributes them throughout most of the extracellular space where they effectively protect the fish from freezing at temperatures as low as the freezing temperature of seawater (-1.8°C) (Hew and Fletcher, 1985; Davies et al., 1988).

Characterization of Antifreeze Polypeptides

Antifreeze proteins have the unusual property of lowering the freezing temperature but not the melting temperature of the serum. The difference between the freezing and melting temperature, commonly known as thermal hysteresis, is a measurement of the presence of macromolecular antifreeze activity. Thermal hysteretic activity is independent of the presence of inorganic salts such as NaCl and can be measured easily with a nanolitre osmometer (Kao et al., 1986). Using this method, we have been able to identify antifreeze proteins in sera of several Newfoundland fishes.

Protein fractionation and characterization of these proteins has led to their classification into antifreeze glycoproteins (AFGP) and antifreeze protein/peptides (AFP). The AFP are quite diverse in their protein structure and can be further divided into three distinct types (I, II, and III) based on their protein sequences (Table 1) (Hew and Fletcher, 1985; Davies and Hew, 1990).

Mechanism of Action of Antifreeze Proteins

Regardless of their chemical structure, the mode of action of the AFGP/AFP is similar. They lower the freezing temperature via a noncolligative mechanism by binding to specific ice crystal planes thereby inhibiting further growth (for review, see DeVries, 1984; Davies and Hew, 1990; Hew and Yang, 1992). Specific binding faces have also been observed by Knight et al., (1990).

The winter flounder type I AFP is the only antifreeze that has been structurally characterized by x-ray crystallography (Yang et al., 1988). Therefore only its mode of action will be discussed.

The winter flounder *Pseudopleuronectus americanus* produces several type I AFP during the winter months. The two major components, HPLC-6 and HPLC-8 together represent 60% of all the AFP in the serum (Fourney et al., 1984a). Both polypeptides have 37 amino acids of which the C-terminal residue, arginine, is amidated. These peptides are rich in alanine (60 mole %) and contain three 11 amino acid repeats of Thr $(X)_2$ Asp/Asn $(X)_7$ where X is mostly alanine. HPLC-8 differs from HPLC-6 in three positions (the amino acid substitutions from HPLC-6 to HPLC-8 are, Lys_{18}-->Ala, Glu_{22}-->Lys and Ala_{26}-->Asp). This results in the position of the salt bridge being shifted towards the C-terminal. The structure of the major AFP in yellowtail flounder, differs from that of the winter flounder in that it contains four 11 amino acid repeats and two salt bridges (Fig. 1). Both circular dichroism and x-ray studies have demonstrated that the polypeptides are α-helical with amphiphilic characteristics (Fig. 1), where most of the hydrophilic amino acids are aligned on one side of the helix and presumably represent the ice-binding motifs (for review, see Yang et al., 1988; Davies and Hew, 1990). Earlier hypotheses suggest that these hydrophilic side chains hydrogen bond with the ice and retard its normal growth (DeVries and Lin, 1977). To further explain the preferential binding of the AFP to the prism face of an ice crystal, Yang et al., (1988), postulated an interaction of the macromolecular peptide dipoles with ice (Fig. 2). It was argued that the dipole field of the alpha helix would align with the dipole of individual water molecules in the ice crystal, thus inducing a dipole-dipole interaction between the protein molecule and the ice crystal. It was observed that the direction for maximum dipole induction would be along the <221> vectors. Since these vectors can be found on the prism face but not on the basal plane,

Table 1.

Structural characteristics of antifreeze proteins.

	AFGP	AFP Type I	AFP Type II	AFP Type III
Carbohydrate	Yes	No	No	No
Mass	2,600 to 33,000	3,300 to 4,500	11,000 to 24,000	6500
Primary structure	(Ala Ala Thr)n disaccaride	alanine-rich multiple of eleven aa repeats	cystine-rich disulfide linked	general
Secondary structure	expanded	α helical amphiphilic	ß sheet	structured but not distinct
Tertiary structure	N.D.	100% helix	N.D.	N.D.
Biosynthetic precursor	multiprotein	prepro AFP	prepro AFP (?)	pro AFP
Protein components	8	7	2-6	12
Gene copies	N.D.	40-50	15	150
Fish species	antarctic nototheniids	right eyed flounders (winter flounder)	sea raven	eel pouts
	northern cods	Sculpins (shorthorn)	smelt	(ocean pout, wolffish)
			herring	

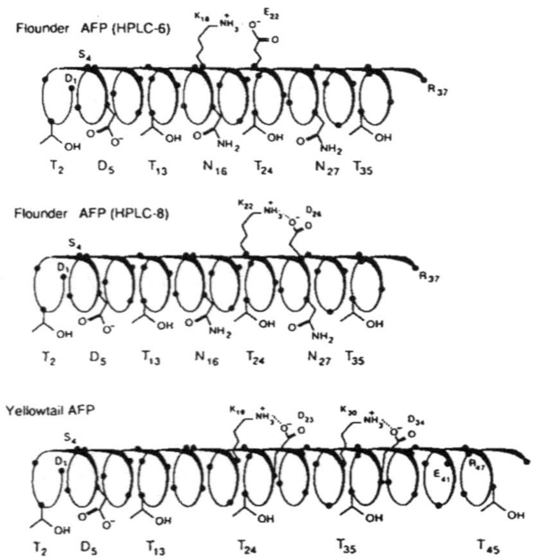

Fig. 1. Secondary structure of type I AFP from righteye flounder.

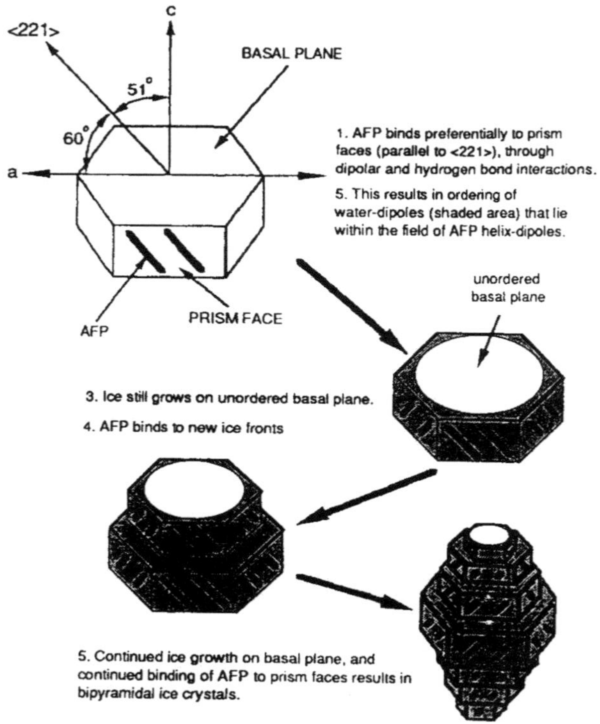

Fig. 2. Schematic representation of type I AFP interaction with ice.

the dipole-induced dipole interaction then confers the "specificity" for prism face binding. Due to the structural diversity of different antifreezes, this dipolar mechanism may not be applicable to other types of antifreezes where more structural work is needed.

Antifreeze Production in Winter Flounder

Winter flounder inhabit the shallow inshore coastal waters of eastern Canada throughout the winter. Therefore they are an excellent model fish by which to study mechanisms of freezing resistance.

In Newfoundland populations of winter flounder, mature AFP appear in the blood plasma in November when the water temperature approximates 4 to 6°C, reach peak values by January, and disappear during May when the temperature begins to rise above 0°C (Fig. 3) (Fletcher, 1977). These seasonal changes in plasma AFP levels are accompanied by concomitant changes in the winter flounder's resistance to freezing. Moreover, the winter increase in freezing resistance over that observed during the summer is identical to the winter increase in plasma antifreeze activity (Fig. 4) (Fletcher et al., 1986).

The environmental and physiological factors regulating the annual AFP cycle in winter flounder have been reviewed in detail by Davies et al., (1988) and Fletcher et al., (1989). All of the available data suggest that the annual cycle is endogenous, with photoperiod acting via the pituitary gland to regulate the precise time of onset of AFP production in the fall (Fletcher, 1981; Fourney et al., 1984 a, b). The pituitary factor responsible for this regulation has been identified as growth hormone (Idler et al., 1989).

Water temperature does not appear to play a major role in controlling the time of AFP production, but it does affect the rate of AFP disappearance from the blood (Vaisius et al., 1989). At the water temperatures that prevail in Newfoundland in the spring (March and April) (0 to -1°C) plasma AFP levels remain at winter values (February) despite the fact that AFP synthesis has stopped. Only when the water temperatures increase above 0°C will AFP levels decline (Fletcher, 1981; Hew et al., 1986). This means that the fish are protected from freezing even when there is an unusally prolonged winter.

Antifreeze Protein Genes

In the past several years, both the cDNA and genomic structure of all the AFP types have been characterized (for review, see Davies et al., 1989). More recently, the gene structure of AFGP of an Antarctic nototheniid has also been reported (Hsiao et al., 1990).

All of the AFP belong to multi gene families. There are approximately 12-15 copies of the gene in sea raven (type II), 40-50 copies in winter flounder (type I) and 150 copies in ocean pout (type III). Many of the gene copies are tandemly repeated as gene clusters (Davies et al., 1989). In the case of AFGP, the genomic clone consists of 46 tandem repeat segments, each encoding one AFGP peptide plus a 3 amino acid space. Most of the repeats (44 copies) represent AFGP-8 and 2 copies of AFGP-7 plus a potential signal peptide. The individual AFGP can be released from the multiprotein presumably using chymotrypsin-like proteases (Hsiao et al., 1990). It appears that multiple gene copies or the processing of the multiproteins are the two preferred mechanisms adopted by various species of fish to increase the protein concentration in the blood needed for freeze-resistance.

The gene structure of flounder AFP is shown in Figure 5. It has a single intron of 600 bp within a 1kb gene sequence (Scott et al., 1985). The intron also contains several liver-specific cis-acting DNA sequences (LF-I like) similar to those seen in other mammalian liver-specific genes such as albumin and antitrypsin (Chan and Hew, unpublished).

Atlantic Salmon Aquaculture: The Importance of Freeze-Resistance

The east coast of Canada is characterized by sub-zero seawater temperatures and the predominance of ice during the winter months (Fig. 6). Since Atlantic salmon and other salmonids do not possess antifreeze peptides their freezing temperatures are approximately -0.7°C. Therefore, sea cage culture is restricted to the most southerly coastal regions, and is largely concentrated in, and around, Passamaquoddy Bay, New Brunswick where water temperatures rarely decline below 0°C (Fig. 4).

With the current rapid decline in wild fish stocks, aquaculture of salmon and other fish species is becoming increasingly important to the socio-economic well being of the many communities scattered along most of the eastern shoreline. However, unless brood stocks of more freeze-resistant fish can be developed, sea-cage culture in most areas will be impossible.

Fig. 3. Annual cycles of plasma freezing temperatures and plasma antifreeze peptide (AFP) levels in Newfoundland winter flounder. (All values plotted as means ± one standard error (N=5 to 10)).

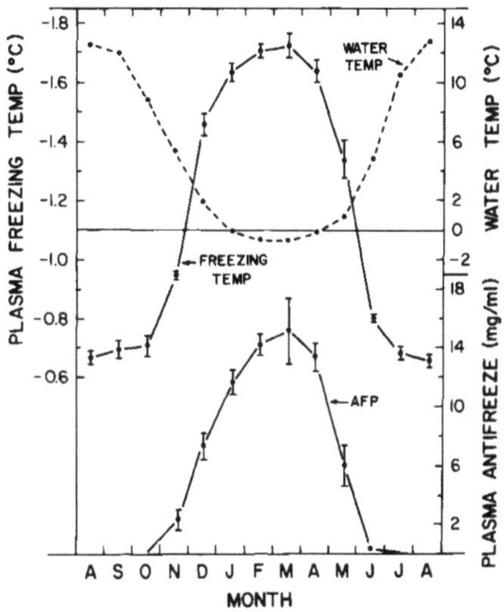

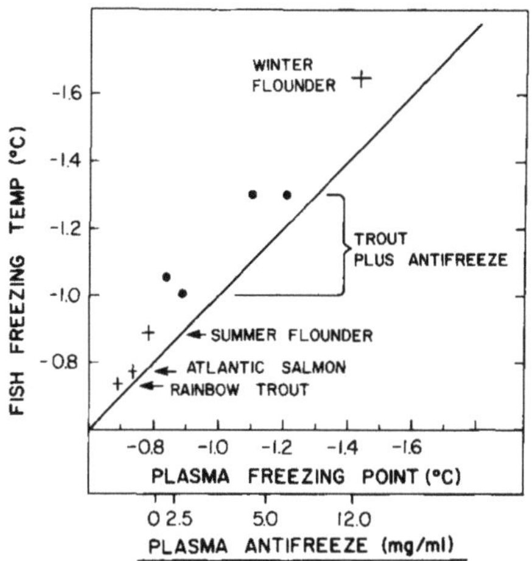

Fig. 4. Relationship between lethal freezing temperatures and plasma freezing temperatures. (Lethal freezing temperatures were determined on seawater-adapted fish by gradually lowering the temperature of the aquaria with the use of crushed seawater ice. "Winter" flounder are winter flounder acclimatized to winter conditions. "Summer" flounder are winter flounder acclimated to summer conditions. Experiments on salmon and rainbow trout were carried out during late winter. Crosses represent means ± one standard error. Solid dots represent individual samples from rainbow trout injected with winter flounder AFP. The diagonal line indicates the water temperature at which the fish should die as predicted from the plasma freezing points. (Fletcher et al., 1986, 1988a).

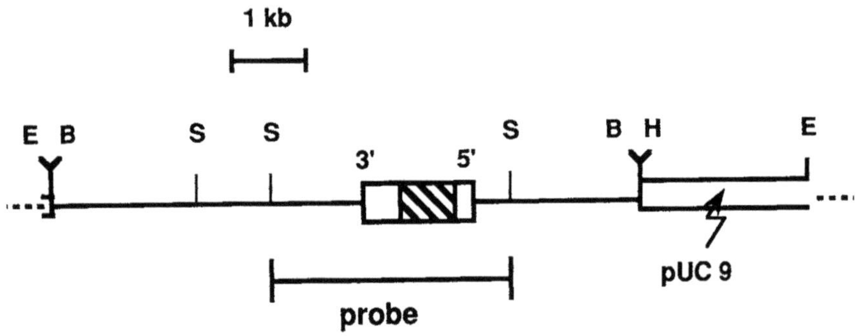

Fig. 5. A schematic diagram of flounder AFP gene (2A-7) used in antifreeze gene transfer in Atlantic salmon. Plasmid 2A-7 linearized at its unique Eco RI site (E), is shown integrated into salmon genomic DNA (broken line). The pUC-9 section of 2A-7 is indicated by double lines and the flounder DNA insert by a single line within which is located a rectangular box representing the 1-kb AFP gene. The open areas of the box are the exons and the hatched area is the intervening sequence. The cleavage sites of Bam HI (B), Sst I (S), and Hind III (H) are marked. The 2.7-kb Sst I fragment used as a probe is underlined. From Davies et al. (1990).

In 1982, with a grant from the Canadian Natural Sciences and Engineering Research Council (NSERC) we elected to attempt to produce more freeze-resistant salmon through the transfer of antifreeze protein genes. At that time the only antifreeze gene we had isolated was from the winter flounder. Therefore most of our transgenic research has focused on the use of this gene.

Can Antifreeze Peptides Confer Freezing Resistance to Fish?

Because we were embarking on a project to improve the freezing resistance of Atlantic salmon by winter flounder gene transfer, we needed direct evidence that the AFP was not species specific and that it would increase the freezing resistance of salmon.

The lethal freezing temperature of both seawater-acclimated Atlantic salmon and rainbow trout is approximately -0.75°C (Figure 4) (Fletcher et al., 1988a). When AFP purified from the blood plasma of winter flounder were injected into the rainbow trout,

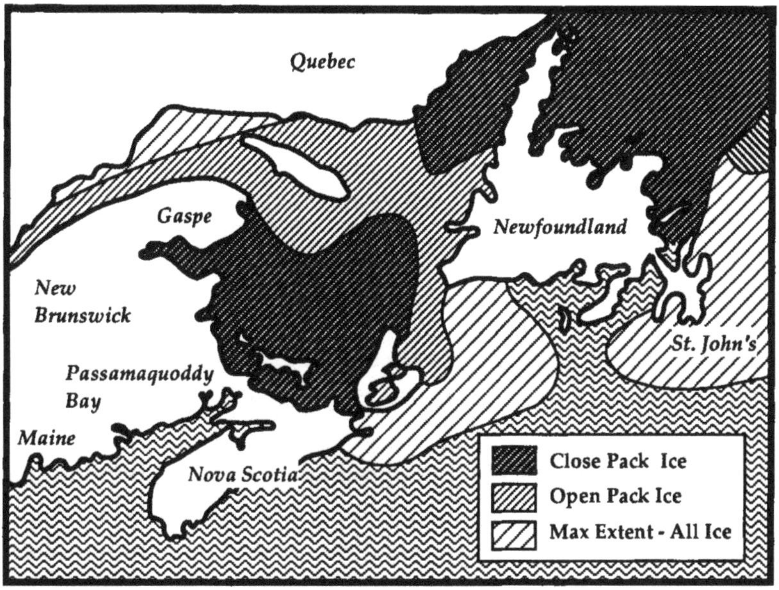

Fig. 6. Average spring ice conditions in Atlantic Canada (adapted from Cook et al., 1987). Brief periods (days-weeks) of sub-zero water temperatures and surface ice can occur at any time during the winter throughout the region.

their freezing resistance increased in direct proportion to blood AFP levels (Figure 4) (Fletcher et al., 1986). These results indicate that Atlantic salmon can be engineered to tolerate subzero water temperatures and ice if winter flounder AFP genes can be inserted into the salmon genome and expressed in physiologically significant quantities.

Antifreeze Protein Gene Transfer

From the outset of our studies we have concentrated on microinjection to transfer AFP genes into salmon eggs. Details of our strategy and techniques are presented in Shears et al., 1992, (see Chapter 3, this monograph). Briefly, approximately 1 x 10^6 copies of linearized winter flounder AFP gene linked to pUC-9 were injected into fertilized salmon eggs through the micropyle (Fletcher et al., 1988b).

Evidence for AFP Gene Transfer to Atlantic Salmon

The results presented here are from experiments that were initiated during the fall spawning period of 1985. Eight months after injection, genomic DNA was extracted from thirty fingerlings (1-2g each) for Southern blot analysis. DNA from two fish gave positive hybridization signals when probed with a labelled AFP gene sequence. The restriction pattern shown by the DNA was consistent with that of the injected DNA, and indicated that it was integrated into the salmon chromosomal DNA (Fletcher et al., 1988b).

Further evidence for integration of the AFP gene was obtained three years after the initial injection, when sperm, from a precocious male which had earlier been shown to possess the AFP gene by Southern blotting, proved postive in the same test (Davies et al., 1990).

More recently, the simplicity of assaying whole blood samples for transgenesis by PCR has made it feasible to screen all surviving fish from the 1985 injection series for the flounder AFP gene (see Davies et al., 1992, chapter 4, this monograph for PCR strategies and techniques). Out of the 324 salmon screened, 10 were transgenic for the antifreeze protein gene. Thus the overall success rate for AFP gene transfer was approximately 3%.

Inheritance of the Flounder AFP Gene in Transgenic Salmon

A summary of the results from the inheritance studies carried out are presented in Table 2. The first opportunity to test for inheritance of the AFP transgene came in the fall of 1988 when one of the male transgenic salmon became sexually mature (Shears et al., 1991). When this male (1985-P_1-1) was crossed with a normal, wild-type female only 17% of the offspring were positive for the presence of the flounder AFP gene. In the fall of 1989 the same male was crossed with another wild-type female resulting in a similar result of 15% of the progeny containing the AFP gene. At the same time 64% of the progeny from a cross between another male (1985-P_1-2) and a wild type female contained the AFP gene. It is evident from these analyses that the two transgenic male parents were germ-line mosaics for the incorporation of the AFP gene. These results are in agreement with the literature on transgenic fish indicating that mosaicism is extremely common in fish developed from injected eggs (Fletcher and Davies, 1991).

During the fall of 1990 four transgenic F_1 males derived from the fall 1988 mating of male 1985-F_0-1 and a wild type female were back crossed to wild-type females. Analyses of hatched embryos (alevins) by PCR showed that in all cases approximately 50% were positive for the presence of the flounder AFP gene. This percentage is what would be expected from Mendelian inheritance of a transgene integrated into a single chromosomal site of the F_1 male parent.

These results clearly indicate the establishment of stable lines of Atlantic salmon transgenic for the winter flounder AFP gene (Shears et al., 1991).

Expression of AFP Genes in Transgenic Salmon

The blood plasma of potential transgenic salmon was screened for the presence of AFP using immunoblotting (Western) procedures. The results obtained from screening all of the salmon remaining from the 1985 injection series indicated that 40% of those that were transgenic for the AFP gene contained low but detectable levels of winter flounder AFP (Davies et al., 1990). However the molecular mass of the presumptive AFP was approximately 6kd, twice that of the mature AFP found in winter flounder (3.3kd).

An explanation for this observation comes from our knowledge of AFP gene expression in winter flounder. The winter flounder AFP gene used in this study codes for an 82 amino acid pre-proAFP (Figure 7) (Pickett et al., 1984). The 23 amino acid

Table 2: Inheritance and expression of the antifreeze protein gene in Atlantic salmon.

Year	Parents		Progeny tested	Transgenic	Expression
	male	female			
1985	wild	wild	324	3%	1%
1988	1985-P_1-1	wild	137	17%	17%
1989	1985-P_1-1	wild	100	15%	---
1989	1985-P_1-1	1985-P_1-2	100	33%	---
1989	1985-P_1-3	wild	100	64%	---
1990	1988-F_1-1	wild	85	53%	---
1990	1988-F_1-3	wild	78	51%	---
1990	1988-F_1-4	wild	85	54%	---

Salmon eggs were injected with the AFP gene in 1985. Data summarized from Shears et al., 1991. AFP gene expression was determined using immunoblotting procedures on serum samples.

10
MET-ALA-LEU-SER-LEU-PHE-THR-VAL-GLY-GLN-LEU-ILE-PHE-LEU-PHE-
20 30
TRP-THR-MET-ARG-ILE-THR-GLU-ALA-Ser-Pro-Asp-Pro-Ala-Ala-Lys-Ala-Ala-
40 50
Pro-Ala-Ala-Ala-Ala-Ala-Pro-Ala-Ala-Ala-Ala-Pro-Asp-Thr-Ala-Ser-Asp-Ala-Ala-Ala-
60 70
Ala-Ala-Ala-Leu-Thr-Ala-Ala-Asn-Ala-Ala-Ala-Ala-Ala-Lys-Leu-Thr-Ala-Asp-Asn-Ala-
80
Ala-Ala-Ala-Ala-Ala-Ala-Thr-Ala-Arg-Gly

Fig. 7. Amino acid sequence of the preproprotein precursor to the winter flounder AFP. The precursor is shown with the signal peptide in capital letters and the "pro" region underlined. The C-terminal glycine is also removed during processing.

pre-sequence is cleaved off cotranslationally and the proAFP is secreted into the blood where the pro-sequence is removed within 24 hours to yield a 37 amino acid mature AFP (Hew et al., 1986). The molecular size of the AFP-immunoreactive protein found in salmon plasma is essentially the same as that of the pro-AFP (6kd) isolated from flounder. This suggests that, although the Atlantic salmon can properly express the inserted winter flounder gene and secrete the proAFP into the blood, it lacks the necessary enzyme systems to process the pro-AFP to mature AFP. A similar result has been obtained in transgenetic *Drosophila melanogaster* for which only proAFP was found in the hemolymph (Rancourt et al., 1987).

Studies by Hew et al., (1986) have shown that proAFP isolated from winter flounder liver possesses 60-70% of the antifreeze activity of mature AFP. Thus, although proAFP is less effective than mature AFP at depressing the freezing temperature, it would, nevertheless, confer increased freezing resistance to salmon.

The progeny from a cross between transgenic male 1985-P_1-1 and a wild-type female were also screened for antifreeze expression. All of the salmon that were transgenic for the AFP gene showed evidence of antifreeze proteins in their blood sera (Fig 8, Table 2). The concentration of AFP in these transgenics ranged from approximately 0.07 to 20 ng/ml. This level of expression is approximately 100 fold less than what is needed to provide adequate freezing resistance to salmon if they are to overwinter in ice-laden sub-zero waters.

Despite the relatively low levels of expression exhibited by the antifreeze transgenes it is apparent that AFP gene expression is as heritable as the genes themselves.

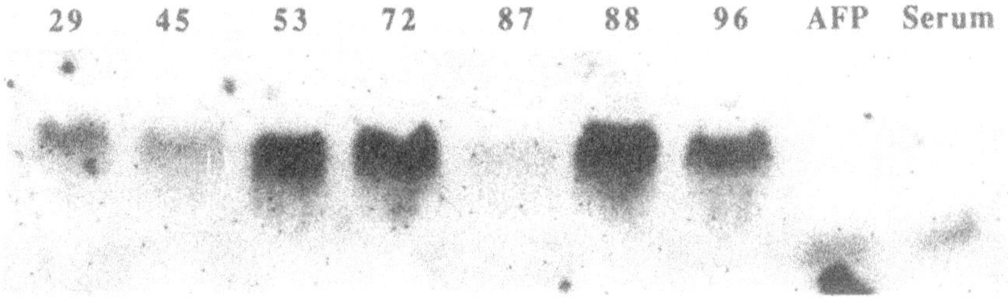

Fig. 8. Expression of proAFP in F_1 transgenic salmon by immunoblot. 3.5 µl of sera from transgenic F_1 #29, #45, #53, #72, #87, #96 were applied, AFP standard (13 ng) and 200 fold dilution of a winter flounder serum. After electrophoretic transfer, the immunoblot was carried out using specific antiflounder AFP antibodies and ^{125}I labelled protein.

Can AFP Gene Expression in Salmon be Improved?

The results of our studies clearly indicate that the winter flounder antifreeze gene has been stably integrated into the Atlantic salmon genome and expressed. We now have to tackle the problem of finding ways to increase the level of antifreeze expression in order to significantly improve the fish's freeze-resistance. Several approaches are being used to achieve this objective.

1. Switch to using a type III AFP gene

Type III AFP does not have a proprotein precursor and is therefore, fully functional after cotranslational cleavage of the signal sequence. Indeed, type III AFP is slightly more active on a molar basis than type I AFP and about twice as active as type I proAFP (Kao et al., 1986). Another reason to switch to the type III AFP is that they appear to be constituitively expressed and are thus not as subject to seasonal regulation as are the type I AFP (Fletcher et al., 1985 a, b).

2. Increase gene dosage

Our research on natural populations of fish has revealed a correlation between gene dosage and the levels of plasma AFP observed during the winter (Scott et al., 1988; Hew et al., 1988). In attempts to mimic the situation in these fish several copies of the AFP genes can be linked in tandem arrays prior to microinjection.

A genetic approach to increasing gene dosage has been adopted using transgenic *Drosophila* lines which has the hsp70-AFP gene locus and, by doing the appropriate genetic crosses it was possible to assay AFP levels in flies bearing one to six copies of the gene. Under heatshock conditions the amount of gene product produced appears to be additive (Rancourt et al., unpublished). The same type of breeding program can be envisioned for salmon which have a large number of small chromosomes.

Another method of increasing the number of antifreeze genes in the salmon is to utilize the AFGP gene recently isolated by Hsiao et al., (1990). This gene encodes a multiprotein that is believed to be cleaved into individual AFGP post-translationally. Thus the transfer of a single AFGP gene could result in multiple copies of AFGP being produced. However, the ability of the salmon to glycosylate the antifreeze nascent chains properly will need to be examined.

3. Use of a stronger promoter/enhancer

The observation that each of the AFP types is encoded by a large multigene family implies that the demand for high concentrations of circulating AFP has been met primarily by gene amplification rather than by strengthening the promoter/enhancer of a single copy gene (Scott et al., 1985, 1986; Hew et al., 1988). The logic of using a stronger promoter has been demonstrated using transgenic *Drosophila* in which two type III AFP genes have been linked to a central yolk protein (YPI,2) promoter. Mature females from these transgenic lines containing just the two genes produce AFP concentrations in their hemolymph in excess of 1 mg/ml (Rancourt et al., 1990).

4. Design a more active and stable AFP

It is possible to produce a superior antifreeze by protein engineering (Mueller et al., 1991; Li et al., 1991). These modifications could involve increased protein chain length and/or increases in the numbers of hydrophilic side chains capable of ice binding. It should also be possible to design AFP that are resistant to in vivo degradation mechanisms so that the AFP can remain in the blood circulation for longer periods of time. In our own studies of the type III AFP from ocean pout, the N-terminal addition of Met Lys and the retention of C-terminal Lysine to produce Met Lys AFP Lys is sufficient to produce a more active and thermally more stable antifreeze (Li et al., 1991). With the elucidation of the 3 dimensional structure of the other AFPs, this approach may offer a systematic and rational design for producing a more active antifreeze.

ACKNOWLEDGMENTS

We would like to extend our appreciation to members of our research team, Madonna King, Sherry Gauthier, Sally Goddard, Margaret Shears, Zhiyuan Gong and Ming Kao who make our contribution to this field possible. The research reported in this manuscript is supported by grants from the Natural Sciences and Engineering Research Council of Canada, the Medical Research Council of Canada and the Canadian Department of Fisheries and Oceans.

REFERENCES

Cook, R. H., Sander, F. and Drinnan, R. E. (1987) Atlantic marine water resources: their utilization and potential for aquaculture. Aquaculture Association of Canada, 4th Annual Meeting Hotel Loews Le Concorde, April 29-May1, 1987. ISBN 2-551-06801-0, 61-86.

Davies, P. L., Hew, C. L. and Fletcher, G. L. (1988) Fish antifreeze proteins: Physiology and evolutionary biology. Can. J. Zool. 66: 2611-2617.

Davies, P. L., Fletcher, G. L. and Hew, C. L. (1989) Fish antifreeze protein genes and their use in transgenic studies. In: Oxford Surveys on Eukaryotic Genes, N. Maclean (ed), Oxford University Press, 6: 85-109.

Davies, P. L. and Hew, C. L. (1990) Biochemistry of fish antifreeze proteins. FASEB. J. 4: 2460-2468.

Davies, P. L., Hew, C. L., Shears, M. A. and Fletcher, G. L. (1990) Antifreeze protein expression in transgenic salmon. In: Transgenic models in medicine and agriculture. R. B. Church (ed), Alan R. Liss, pp.141-161.

Davies, P. L. and Gauthier, S. Y. (1991) The application of PCR to aquaculture. In: Transgenic fish. C. L. Hew and G. L. Fletcher (eds.) World Scientific Publishing, Singapore, pp. 61-71.

DeVries, A. L., Komatsu, S. K. and Feeney, R. E. (1970) Chemical and Physical properties of freezing point depressing glycoproteins from Antarctic fishes. J. Biol. Chem. 245: 2901-2913.

DeVries, A. L., and Lin, Y. (1977) Structure of a peptide antifreeze and mechanism of adsorption to ice. Biochem. Biophys. Acta. 495: 380-392.

DeVries, A. L. (1984) Role of glycopeptides and peptides in inhibition of crystallization of water in polar fishes. Phil. Trans. R. Soc. Lond. B 304: 575-588.

Fletcher, G. L. (1977) Circannual cycles of blood plasma freezing point and Na^+ and Cl^- concentrations in Newfoundland winter flounder *(Pseudopleuronectes americanus)*: Correlation with water temperature and photoperiod. Can. J. Zool. 55: 789-795.

Fletcher, G. L. (1981) Effects of temperature and photoperiod on the plasma freezing point depression, Cl^- concentration, and protein "antifreeze" in winter flounder. Can. J. Zool. 59: 193-201.

Fletcher, G. L., Haya, K., King, M. J. and Reisman, H. M (1985a) Annual antifreeze cycles in Newfoundland, New Brunswick and Long Island winter flounder (*Pseudopleuronectes americanus*). Mar. Ecol. Prog. Ser. 21: 205-212.

Fletcher, G. L., Hew, C. L., Li, X. M., Haya, K., and Kao, M. H. (1985b) Year-round presence of high levels of plasma antifreeze peptides in a temperate fish, ocean pout (*Macrozoarces americanus*). Can. J. Zool. 63, 488-493.

Fletcher, G. L., Kao, M. H. and Fourney, R. M. (1986) Antifreeze peptides confer freezing resistance to fish. Can. J. Zool. 64: 1897-1901.

Fletcher, G. L., Kao, M. H. and Dempson, J. B. (1988a) Lethal freezing temperatures of Arctic char and other salmonids in the presence of ice. Aquaculture. 71: 369-378.

Fletcher, G. L., Shears, M. A., King, M. J., Davies, P. L. and Hew, C. L. (1988b) Evidence for antifreeze protein gene transfer in Atlantic salmon (*Salmo Salar*). Can. J. Fish. Aquat. Sci. 45, 352-357.

Fletcher, G. L., Idler, D. R., Vaisius, A. and Hew, C. L. (1989) Hormonal regulation of antifreeze protein gene expression in winter flounder. Fish. Physiol. Biochem. 7: 387-393.

Fletcher, G. L. and Davies, P. L (1991) Transgenic fish for aquaculture. In: "Genetic engineering principles and methods." J. K. Setlow (ed). Plenum Press, 13: 331-370.

Fourney, R. M., Fletcher, G. L. and Hew, C. L. (1984a) The effects of long day length on liver antifreeze mRNA in the winter flounder (*Pseudopleuronectes americanus*). Can. J. Zool. 62: 1456-1460.

Fourney, R. M. Fletcher, G. L. and Hew, C. L. (1984b) Accumulation of winter flounder antifreeze messenger RNA after hypophysectomy. Gen. Comp. Endocrinol. 54: 392-401.

Gordon, M. S., Andur, B. N. and Scholander, P. F. (1962) Freezing resistance in some northern fishes. Biol. Bull. 122: 52-62.

Hew, C. L., and Fletcher, G. L. (1985) Biochemical adaptation to the freezing environment-structure, biosynthesis and regulation of fish antifreeze polypeptides. In: Circulation, Respiration and Metabolism. R. Gilles (ed). Springer-Verlag, Berlin Heidelberg. 553-563.

Hew, C. L., Wang, N. C., Yan, S. Cai, H., Sclater, A. and Fletcher, G. L. (1986) Biosynthesis of antifreeze polypeptides in winter flounder: Characterization and seasonal occurrence of precursor polypeptides. Eur. J. Biochem. 160: 267-272.

Hew, C. L., Wang, N.C., Joshi, S., Fletcher, G. L., Scott, G. K., Hayes, P. H., Buettner, B. and Davies, P. L. (1988) Mutiple genes provide the basis for antifreeze protein diversity and dosage in the ocean pout, *Macrozoarces americanus*. J. Biol. Chem., 263, 12049-12055

Hew, C. L. and Yang, D. S. C. (1992) Protein interaction with ice. Eur. J. Biochem. (In press).

Hsiao, K. C., Cheng, C. H. C., Fernandes, I. E., Detrich, H. W. and DeVries, A. L. (1990) An antifreeze glycopeptide gene from the Antarctic cod *Notothenia coriiceps neglecta* encodes a polyprotein of high peptide copy number. Proc. Natl. Acad. Sci. USA 87: 9265-9269.

Idler, D. R., Fletcher, G. L., Belkhode, S., King, M. J., and Hwang, S. J. (1989) Regulation of antifreeze protein production in winter flounder: A unique function for growth hormone. Gen. Comp. Endocrinol., 74: 327-334.

Kao, M. H., Fletcher, G. L., Wang, N. C. and Hew, C. L. (1986) The relationship between molecular weight and antifreeze polypeptide activity in marine fish. Can. J. Zool. 64: 578-582.

Knight, C. A., Cheng, C. C. and DeVries, A. L. (1990) Adsorption of α-helical antifreeze peptides on specific ice crystal surface planes. Biophys. J. 59: 409-418.

Li, X., Trinh, K. Y. and Hew, C. H. (1991) Expression and characterization of an active and thermally more stable recombinant antifreeze polypeptide from ocean pout, *Macrozoarces americanus*, in *E. coli*.- Improved expression by the modification of the secondary structure of the mRNA. Protein Engineering. (In press).

Mueller, G. M., McKown, R. L., Corott, L. V., Hague, C. and Warren, G. J. (1991) Inhibition of recrystallization in ice by chimeric proteins containing antifreeze domains. J. Biol. Chem. 266: 7339-7344.

Pickett, M., Scott, G., Davies, P. L., Wang, N., Joshi, S. and Hew, C. L. (1984) Sequence of an antifreeze protein precursor. Eur. J. Biochem. 143: 35-38.

Rancourt, D. E., Walker, V. K. and Davies, P. L. (1987) Flounder antifreeze protein synthesis under heat shock control in transgenic *Drosophila melanogaster*. Mol. Biol. 7:2188-2195.

Rancourt, D. E. Peters, I. D., Walker, V. K. and Davies, P. L. (1990) Wolffish antifreeze proteins from *Drosophila*. Bio/Technology 8: 453-457.

Scholander, P. F., van Dam, L., Kanwisher, J. W., Hammel, H. T. and Gordon, M. S. (1957) Supercooling and osmoregulation in Arctic fish. J. Cell. Comp. Physiol.49: 5-24.

Scott, G. K., Hew, C. L. and Davies, P. L. (1985) Antifreeze protein genes are tandemly linked and clustered in the genome of the winter flounder. Proc. Natl. Acad. Sci. USA 82: 2613-2617.

Scott, G. K., Fletcher, G. L. and Davies, P. L. (1986) Fish antifreeze proteins: recent gene evolution. Can. J. Fish. Aquat. Sci. 43: 1028-1034.

Scott, G. K., Davies, P. L., Kao, M. H. and Fletcher, G. L. (1988) Differential amplification of antifreeze protein genes in the pleuronectinae. J. Mol. Evol. 27: 29-35.

Shears, M. A., Fletcher, G. L., Hew, C. L., Gauthier, S. and Davies, P. L. (1991) Transfer, expression, and stable inheritance of antifreeze protein genes in Atlantic salmon (*Salmo salar*). Mol. Marine Biol. Biotech. 1: 58-63.

Shears, M. A., King. M. J., Goddard, S. V. and Fletcher, G. L. (1992) Gene transfer in salmonids by injection through the micropyle. In: Transgenic Fish. C. L. Hew and G. L. Fletcher (eds). World Scientific Publishing Co. Singapore, pp. 44-60.

Vaisius, A., Martin-Kearley, J. and Fletcher, G. L. (1989) Antifreeze protein gene transcription in winter flounder is not responsible to termperature. Cell. Mol. Biol. 35: 547-554.

Yang, D. S. C., Sax, M., Chakrabartty, A. and Hew, C. L. (1988) Crystal structure of an antifreeze polypeptide and its mechanistic implications. Nature (London). 333: 232-237.

CHAPTER 12

Ecological and Regulatory Uncertainties Associated with Transgenic Fish

Eric M. Hallerman and Anne R. Kapuscinski

ABSTRACT

The development of transgenic fishes introduces ecological and public policy uncertainties which may constrain the rate of progress towards practical utilization. Pathways giving rise to ecological impacts of transgenic fishes in natural communities may prove complex, and the magnitude of such impacts difficult to predict. Continuing uncertainty over policies regulating environmental release and commercial use has slowed the pace of development of genetically modified organisms. Knowledge of the benefits and risks of using transgenic fishes, effective exercise of regulatory authority and public confidence in each of these will prove important in determining whether transgenic fishes reach practical utilization.

INTRODUCTION

Gene transfer in fish has advanced rapidly, based on the results of numerous studies addressing technical aspects of gene transfer, regulation of transgene expression, and phenotypic alteration due to transgene activity. With the apparent realization of phenotypic alterations (such as accelerated growth) which are attractive for aquaculture or for stocking programs supporting capture fisheries, commercial use of transgenic lines of fish is being contemplated.

Rapid development and commercial use of transgenic fishes will not, however, depend on technical capabilities alone. Development, field testing and commercial use of transgenic lines of fish introduce a range of ecological and public policy uncertainties which must be resolved before transgenic fish can reach practical utilization. The purposes of this review are: to identify ecological and public policy uncertainties facing the development and commercialization of transgenic lines, to describe recent public policy developments addressing these uncertainties, and to suggest pro-active approaches toward resolving such uncertainties. These problems will be discussed within the context of laboratory production, development and field testing, and distribution and commercial use of transgenic fishes. The regulatory system in force in the U.S. will be highlighted because it is relatively well developed and may serve as a model for other countries. Reference will be made to parallel regulatory systems in other countries.

LABORATORY PRODUCTION OF TRANSGENIC INDIVIDUALS

"The public does not have to understand what is going on in biotechnology ... But people need a perception that somebody is in charge and somebody is looking out for their best interests (W. Hyer, cited in Office of Technology Assessment 1988, page 43)."

Transgenic individuals are produced in the laboratory through microinjection of a solution bearing a recombinant DNA construct into newly fertilized fish eggs. The young fish are reared indoors and samples of their tissues are screened for presence of the DNA construct and evidence of transgene activity. Individuals proving positive are, when possible, raised indoors through maturity and spawning. With some species, this has proven impossible, and confinement in outdoor facilities is necessary (see section below). Individuals which were transformed in the germ line and which transmit the functional transgene to their offspring, can serve as founders of transgenic lines.

Public policy implications in the U.S.: Laboratory production of transgenic animals has not proven controversial with the general public because:
(1) confidence about the effective regulation of recombinant DNA research has become established, and laboratory, containment is considered complete. These perceptions generally hold true. Close examination of existing public policy and of laboratory practice, however, reveals flaws and inconsistencies.

In the early days of recombinant DNA research, fears over the safety of such work led to the development of guidelines for safe laboratory practices (National Institutes of Health 1984, 1986). With the accumulation of knowledge and experience, some provisions of the guidelines were relaxed. Public fears of recombinant DNA technology subsided.

Controversies over proposed field tests of recombinant DNA-bearing organisms led to development of the Coordinated Framework for the regulation of Biotechnology (Office of Science and Technology Policy 1984, 1985, 1986). Under the Coordinated Framework, production of transgenic animals in institutions that receive support from NIH or from other federal agencies is regulated under the NIH guidelines. Among institutions not receiving NIH or other federal funding, voluntary compliance with the NIH guidelines is "expected." There are no mechanisms for enforcement in all cases.

Commercial development of transgenic animals will be concentrated within the private sector. Extension of provisions of the Coordinated Framework or passage of new legislation will be required to bring commercial production of transgenic organisms under regulatory purview. To our knowledge, no such regulatory initiative is being contemplated.

In institutions where compliance with the NIH guidelines is mandatory, compliance is generally, but not always, complete. Administration of the guidelines is decentralized, the responsibility being delegated to Institutional Biosafety Committees (IBCs). The guidelines are vague about what constitutes effective laboratory containment, leaving much to the judgement of an IBC. In many cases, IBCs lack the necessary expertise to make informed judgements of particular containment facilities or protocols (General Accounting Office 1987). Lapses in recognition of what constitutes effective containment have resulted in situations where, for example, gene transfer experiments have been carried out in laboratories lacking effective screening of effluent water.

Certain new recombinant DNA methodologies are not strictly covered by the NIH guidelines because of problems with the existing definition of genetically engineered organisms (Morgan, 1990). Critics also question the efficacy and direction of the committee administering the guidelines. Furthermore, NIH doesn't want to regulate animals within its purview, and regulation of transgenic animal research will ultimately be transferred to the U.S. Department of Agriculture (USDA). Promulgation of the USDA guidelines has been subject to a long delay.

Regulatory uncertainties notwithstanding, to date no accidental introductions of transgenic fishes into natural environments have become apparent. Concerned scientists should take the measures necessary to ascertain that their laboratories are indeed secure. They might also advocate extension of regulatory authority to all public and private recombinant DNA laboratories, not just those receiving federal funding.

Regulatory structures in other countries: An ad hoc group of experts commissioned by the Organisation for Economic Cooperation and Development (OECD), a grouping of 25 industrialized nations, developed recommendations concerning safety in use of recombinant DNA-bearing organisms in industry, agriculture and the environment. That report (OECD, 1986) has influenced development of such policies worldwide (Ager, 1988). In most OECD member countries, existing legislation related to health or environmental protection is being applied to regulate biotechnology.

There are a number of Canadian federal acts which might regulate the development of biotechnology products (Government of Canada, 1988). Production of transgenic animals is regulated under guidelines promulagated by the Medical Research Council (MRC). Weaknesses in existing regulatory authority (Kapuscinski and Hallerman, 1990c) include the requirement to follow the guidelines only among projects funded by the MRC or by the Natural Sciences and Engineering Research Council (NSERC).

Despite a diversity of existing biotechnology regulations among western European nations, a directive on regulations over contained use of genetically modified organisms was recently adopted by the European Economic Community (EEC) (Dixon, 1990c). The directive sets out minimum containment conditions; member countries can impose stricter rules if they wish. Member countries will have 18 months to harmonize their regulations with the directive (Hodgson, 1990b).

Responsibility for regulating laboratory production of transgenic animals in Japan is divided among several agencies, depending upon where it is carried out (McCormick, 1987): research carried out at universities is subject to guidelines promulgated by the Ministry of Education, at other research organizations to guidelines of the Science and Technology Agency, and in industry to guidelines of the Ministry of International Trade and Industry.

Research on transgenic fish is being carried out in a number of countries where there are no legal regulations mandating safe laboratory practice or restricting environmental release of genetically modified organisms (Table 1). As concerned scientists, we hope that this work is being executed with suitable safeguards and will soon become subject to mandatory safeguards.

ENVIRONMENTAL IMPACTS OF TRANSGENIC FISHES

"The long-term effects of a transgenic mirror carp escape cannot be entirely predicted... However, the anticipated altered performance in growth rate of the transgenic mirror carp does suggest the likelihood of ecological impact on the existing aquatic ecosystem once an escape occurs (Coorperative State Research Service, 1990, page 5765)."

"A natural historian who knows [the] details [of how a species lives, reproduces, disperses, and competes] or knows how to study them, and who appreciates the broad ecological picture may be the best person to call on for information on how a modified organism is likely to behave in its new habitat (Regal, 1986)."

Phenotypic alteration of transgenic fishes: Data from original generation transgenic fish indicate varying degrees of transgene-induced alteration of targeted traits (Kapuscinski and Hallerman, 1990a). At present, there is little or no information on the magnitude of transgene induced direct effects in more advanced generations.

Besides the intended direct effects, indirect phenotypic changes might occur as a consequence of transgene integration or expression or of pleiotropic effects (Kap-

Table 1. Examples of countries where research on gene transfer in fish is being conducted, and presence or absence of regulations over research with or envirorimental release of genetically modified organisms.

Country	Source of regulation[1]		
	National	EEC Member[2]	OECD member[3]
Canada	yes	no	yes
France	yes	yes	yes
Germany	yes	yes	ves
Hungary	no	no	no
India	no	no	no
Indonesia	no	no	no
Israel	no	no	no
Ireland	no	yes	yes
Japan	yes	yes	yes
Malaysia	no	no	no
Norway	yes	no	ves
People's Rep. of China	no	no	no
Thailand	no	no	no
U.S.S.R.	no	no	no
United Kingdom	yes	yes	yes
United States	yes	no	yes

[1] In some cases, there may be regulation of research involving recombinant DNA at particular institutions.

[2] Compliance with European Economic Community (EEC) directives implies prompt enactment of regulations.

[3] Suggests voluntary compliance with recommended Organisation for Economic Cooperation and Development (OECD) guidelines.

uscinski and Hallerman, 1990a). In particular, expression of the trangene could impact a range of physiological processess in the host, giving rise to unexpected or unintended phenotypic alterations. To account for this in evaluating aquaculture performance or anticipating ecological impacts, it will be important to estimate both means and variances for quantitative traits not necessarily targetted by the gene transfer experiment. Traits such as age at sexual maturity, fertility, food conversion, and behavior will influence the performance and fitness of the fish in aquaculture or in the wild.

In terms of the potential ecological impact of a transgenic fish, it is neither the source of the transferred gene nor the presence or absence of a homologous gene in the host's genome which will prove important. Instead, it is the type and magnitude of phenotypic alteration in the transgenic animal (Tiedje et al. 1989; Kapuscinski and Hallerman, 1990a). Although current knowledge does not allow us to predict the full range or magnitude of phenotypic changes in particular lines of transgenic fish, we can draw upon knowledge of fish biology to outline different classes of phenotypic alterations unique to transgenic fishes (Kapuscinski and Hallerman, 1990c):

- Changes in metabolic rates. Exemplified by transfers of growth hormone genes, growth factors affect many processes in the host, including energy partitioning among growth, respiration, reproduction, and osmoregulation in some species.
- Changes in tolerances to physical factors. Exemplified by transfer of antifreeze polypeptide genes, changes in tolerance to physical environmental factors could shift or extend the range of habitats where a species might persist.
- Changes in behavior. For example, transfers of endocrine compounds could affect traits beyond that intended because hormones frequently regulate more than one physiological process in the host.
- Changes in resource use. Transgenes closing a genetic lesion in a key biochemical pathway could eliminate dietary requirements for certain compounds. Factors affecting host energy budgets could affect appetite, prey selection, food conversion, and growth.
- Changes in resistance to parasites, pathogens, or predators. Advances in knowledge and techniques could form the basis for transfer of viral epitope glycoprotein genes or antigen receptor genes.

Ecological impacts of transgenic fishes: Phenotypic changes within these or additional categories would form the basis for ecological impacts of transgenic fishes. The different classes of transgene-induced phenotypic alterations might impact wild populations through a variety of mechanisms. Examples of impact pathways would include the following:

- For transgenes affecting energy partitioning, the large size of individuals bearing introduced growth hormone genes might favor them in agonistic interactions over food resources.
- For transgenes affecting physical tolerances, expression of an antifreeze polypeptide in salmon would encourage expansion of sea cage aquaculture to new areas, which could lead to escape of transgenic individuals. Elevated abundance of non-native salmon in these areas could lead to ecological or genetic impacts on native salmon.
- Because hormones directly and indirectly affect behavior, transfer of such genes could impact the successful execution of life history-related events such as migration, territoriality, and mating.
- Altered resource or substrate use could result if, through accelerated growth, predatory transgenic individuals exhibit a larger mouth gape and make use of new prey items.
- Improved resistance to diseases or parasites, while generally perceived as a positive development, would lead to transgenic individuals being favored in natural communities, contributing to any tendency for them to predominate there and contribute to the decline of natural populations.

Such phenotypic changes would be most important if they affect life history patterns or spatiotemporal use of habitat resources. The likelihood and magnitude of ecological impacts are also dependent upon the effects of phenotypic changes upon the fitness of transgenic individuals, their ability to survive and leave viable offspring in the wild. Such considerations are important in the case of fish and other aquatic organisms, which have not been changed by domestication to the degree that their ability to survive in the wild has been significantly reduced.

Because of the complexity of aquatic communities, it is difficult to predict the mechanisms through which transgenic fishes might effect ecological impacts. Consideration of principles of aquatic ecology has led to insights regarding possible impacts within lake (Kapuscinski and Hallerman,1990a) and stream (Hallerman and Kapuscinski, 1990c) communities.

Impacts of transgenic fish would include, but would not be limited to those associated with impacts of releasing non-transgenic hatchery fish (Hallerman and Kapuscinski, 1990c). Heightened mortality rates, reduced growth rates and other impacts following stocking of hatchery fish result from heightened competition, social stresses and transmission of parasites and pathogens.

Aquatic communities function by transferring nutrients, energy, information or organisms along numerous interconnected pathways. The complexities of these interactions, complicated by indirect effects and time lags, impede our ability to predict the effects of possibly subtle, but nevertheless important ecological perturbations. Thus, it is difficult to anticipate the community-level effects of transgenic fish upon aquatic ecosystems.

Recognizing that most fishes subject to gene transfer experiments are predators, consideration of their possible effects upon an aquatic community is important. The process of predation plays a large role in ordering the structure of aquatic communities, as recognized in "top down" models of energy cycling (Carpenter and Kitchell, 1988) and in designation of certain predators as "keystone" species (Connell, 1975) within communities. Complicating the consideration is the progression of fishes through several trophic roles as they grow larger. The presence of transgenic predators, particularly those growing rapidly, or to large size, could lead to perturbation of the system's normal predatory interactions, and in turn to restructuring of the food web or destabilization of the community.

The likelihood of establishment of transgenic fishes in a given system would clearly be a function of their numbers. The degree of ecological risk posed by transgenic fish would thus be a function of the scale and freqency of introductions into an aquatic system. Because of the scale of introduction implied, deliberate stocking programs would present greater ecological risk than accidental introductions.

The numbers of individuals accidentally escaping into natural systems would be greatly influenced by the degree of confinement effected by the culture system. Parallel spectra associate the degree of confinement in particular culture systems with the attendant degree of ecological risk (Kapuscinski and Hallerman, 1990a). In general, minimum risk is associated with indoor research laboratories, moderate risk with outdoor research facilities, greater risk with commercial aquaculture operations, and maximum risk with natural aquatic ecosystems. The security associated with a particular holding facility is largely a function of barriers to escape of fish via influent or effluent waters, susceptibility of rearing units to flooding or bird predation, and site security measures against vandalism and theft. Placement of a specific confinement facility along the continua should also reflect the ecological vulnerability and value of nearby aquatic ecosystems (Kapuscinski and Hallerman, 1990c).

Impacts of fertile versus sterile transgenic fish: Should transgenic fish enter natural populations and reproduce, the resultant ecological and genetic impacts could be perpetuated. In terms of risk assessment, the problem is one of estimating the fitness of transgenic genotypes in natural ecosystems. Expression of certain transgenes, for example those giving rise to larger size, might enhance the fitness of transgenic fish relative to non-transgenic native fishes (Hallerman and Kapuscinski, 1990c). Lacking data on the fitness of transgenic individuals, we cannot rule out the possibilities of transgenic phenotypes adapting to natural environments through responses to natural selection and wild populations becoming dominated by transgenic genotypes. Nor can we rule out the possibility of transgenic genotypes becoming extinct. Given the potential importance of reproduction-mediated impacts of transgenic fishes, it is critical that this lack in knowledge be addressed before fertile individuals are released into any but well secured confinements.

Use of sterile transgenic fish would limit the potential impacts posed, depending upon how sterilization was achieved (Hallerman and Kapuscinski, 1990c). Among triploid fishes, there remains the possibility of testis maturation and functional sperm production (Benfey et al., 1989). Embryos produced by crossing triploid and diploid salmonids proved inviable (Thorgaard and Allen, 1987), but crosses of triploid and diploid grass carp produced small numbers of viable progeny (Goudie, 1988). Additionally, batch induction of triploidy can yield individuals which are diploid and fertile. Fish sterilized through hormonal means might not mature physiologically.

Sterilization by whatever means would preclude the possibility of perpetuation of transgenic genotypes in recipient ecosystems. Sterility allows the management option of halting further releases and thereby limiting ecological perturbations should ecological impacts be detected. This is especially important because lags in identifying negative impacts are likely. Monitoring the ecological impacts of sterile transgenics through several generations would seem a wise management approach.

Research needs: A number of unquantified factors affect the risks posed by transgenic fish in natural aquatic systems. Directed research in secure facilities quantifying (1) transgene-induced phenotypic alterations and (2) viability and reproductive success of transgenic fish in natural environments would prove valuable in developing ecological risk assessments for transgenic fish. Aquatic mesocosms (Odum, 1984) and aquatic ecosystem simulation modelling may be useful tools for quantfying the fitness of transgenic fish in natural environments. Research demonstrating improved methods for mass sterilization of fish would be a practical contribution toward safe use of transgenic fishes.

DEVELOPMENT AND FIELD TESTING OF TRANSGENIC LINES

"Such is the human condition: The trains start arriving when we are in the early stages of thinking about how to build a railway station (W.A. Anderson, cited in Witt, 1990, page 111) . "

Natural environmental conditions are required for attainment of sexual maturity in certain fishes, and so outdoor release will be required in propagation of transgenic lines of some species (e.g., common carp). Normal culture conditions will be required for field testing of transgenic lines of any fish species.

Outdoor release of transgenic fish has proven highly controversial because outdoor confinement introduces the possibility of escape of transgenic fish, and entry of trangenic fish into natural systems introduces the possibiltiy of ecological impacts. These considerations were brought to the forefront of regulatory debate with the recent release of transgenic carp at Auburn University (Fox, 1989; Cooperative State Research Service, 1990; Hallerman and Kapuscinski, 1990a).

Public policy implications: It was lack of a regulatory policy in the face of plans to release genetically altered microbes in the environment that produced the impetus for development of the Coordinated Framework for the Regulation of Biotechnology. The working premise underlying the Coordinated Framework was that regulatory authorities mandated under existing, statutes could be extended to cover the products of biotechnology. Thus, the Animal and Plant Health Inspection Service (APHIS) of the U.S. Department of Agriculture (USDA) has stated its authority to regulate genetically altered organisms as "biological products" under the Virus, Serum, Toxin Act and as "plant pests" under the Plant Quarantine Act and the Federal Plant Pest Act (Animal and Plant Health Inspection Service, 1987).

Under unpublished draft guidelines (Office of Agricultural Biotechnology, 1990), requests to USDA for a permit for an environmental release are reviewed by the Agricultural Biotechnology Research Advisory Committee (ABRAC) and by outside experts. Confinement facilities and protocols for the proposed release are evaluated and approved if satisfactory. Two aspects of the policy, now in force on an ad hoc basis, are open to criticism:

- Compliance with the regulatory procedure is required only among projects which are funded by USDA. Many, if not most, gene transfer experiments are not funded by USDA.

- There is great uncertainty about regulatory procedure under the proposed guidelines, now under development for at least two years and in at least their fourth draft (August 1990). If an Environmental Impact Statement on the draft guidelines is executed, it will require at least two more years. One leading researcher has complained that he should not begin construction of a containment faciltiy until final guidelines are in place.

Gene transfer researchers have tended to regard environmental release regulations as an impediment to technical progress. In the current U.S. regulatory regime, that attitude is understandable. Upon final promulgation of release regulations, however, adherence to established protocols will provide researchers with a mechanism for securing a release permit while assuring the public that such studies will be conducted safely.

Given the lack of well developed guidelines or regulatory experience regarding transgenic animals, it is impossible to evaluate the use of redefined exclusionary statutes as a means of regulating use of genetically modified organisms. Scientists tend to be less concerned with the form of the regulations as with their content and efficacy.

Two attempts to establish a regulatory mechanism by express legal mandate have been made to date. In 1989, H.R. (House of Representatives) 1557, a bill for a Transgenic Animals Regulatory Reform Act, died without action. The Omnibus Biotechnology Act, H.R. 5312 (described in Fox, 1990), is under consideration as of this writing. Although neither bill addressed all concerns, either would strengthen the present, incomplete regulatory regime.

Regulatory structures in other countries: Under the suggested OECD (1986) guidelines, requests for environmental releases would be subject to case-by-case review of potential risks. A survey of policies now in force among member nations revealed a variety of regulatory regimes.

The regulatory approach taken in Canada parallels that taken in the U.S., applying existing legislation (such as the Canadian Food and Drug Act, the Quarantine Act, and the Animal Disease and Protection Act) to biotechnology. Although the applicability of such legislation to veterinary, biologics, food or drugs seems straightforward, products such as transgenic animals, which are intended for use in the environment, are not well covered (Kapuscinski and Hallerman, 1990b). The National Biotechnology Advisory Committee (1989) urged the federal government to clarify the coverage of biotechnology products under existing legislation. The Canadian Environmental Protection Act may be applied in situations where regulatory coverage under existing legislation is absent or unclear (Government of Canada, 1988). Draft regulations promulgated under the Act, including those for confinement of transgenic animals and for assessing permit applications for environmental releases, are still in development.

Among European Economic Community (EEC) countries, great divergence of public sentiment about genetically modified organisms led to a corresponding divergence of regulations over environmental releases (Newmark, 1987). Nevertheless, after several years of debate, a European Commission (EC) directive aimed at unifying environmental release regulations among EEC member countries was recently approved (Dixon, 1990c). Member countries have 18 months in which to enact the directive at the national level, and are free to adopt stronger rules if they choose (Hodgson, 1990b).

Thus, in March 1990, the Netherlands adopted a framework regulating the environmental use of genetically engineered organisms which closely parallels the EC directive, but which requires public notification of key features of the proposed release (Hodgson, 1990b). Following adoption of the EC directive, Italy, which had no existing release regulations, is first organizing a regulatory authority (Anonymous, 1990). In the U.K., where existing guidelines had called for prior notification of release of recombinant DNA-bearing organisms into the environment (Dixon, 1989), provisions of a bill expected to become law by the end of 1990 require a researcher to provide an environmental safety assessment and to obtain consent from the Health and Safety Executive (Dixon, 1990a).

Among Scandinavian countries (Newmark, 1987), Denmark has banned the release of any recombinant DNA-bearing organism. In Sweden, environmental release is subject to a voluntary notification scheme. However, the Swedish Minister of Agriculture suggested that Scandinavian countries might ban genetic engineering of animals entirely.

In Japan, regulatory authority over agriculture-related environmental releases of genetically modified organisms lies with the Ministry of Agriculture, Forestry and Fisheries under guidelines adopted in 1986 (McCormick, 1987). Regulatory oversight seems more stringent than in the U.S., with stricter containment standards and more detailed review, including individual inspection of research and production facilities.

COMMERCIAL USE OF TRANSGENIC LINES

"Even if you're on the right track, you'll get run over if you just sit there (Will Rogers, cited in Witt, 1990, page 164)."

Present difficulties in obtaining environmental release permits notwithstanding, steady progress in development of transgenic lines seems likely. The first transgenic lines of fish might emerge from development and field testing within the decade, becoming available for practical utilization.

It is difficult to anticipate how transgenic lines will ultimately be commercialized. Different possibilities include (1) direct spin-off of transgenic lines from public facilities such as university agriculture experiment stations, (2) marketing of transgenic lines from specialty agricultural biotechnology firms, or (3) licensing of transgenic broodstock from either of these sources to existing commercial fish seedstock producers. The importance of these alternative pathways to the marketplace will depend in large degree to resolution of key questions in the public policy arena. Inasmuch as the outcome of this process will affect the availability of funding for research and development, it is of interest to those practicing gene transfer in fish.

Public policy implications: 1. Commercial use permits: In the process of developing transgenic lines, laboratory research and well regulated field testing would seem to pose relatively little risk of significant environmental harm. As transgenic fish are ultimately distributed for commercial use, however, they would presumably enter a wide range of less secure confinements. From there, considerable numbers of transgenic individuals might escape into natural environments, giving rise to a more significant likelihood of ecological impacts. Consideration of public policies regulating commercial use of transgenic fishes is therefore of utmost importance.

Under the Coordinated Framework, regulation of commercial use of genetically modified organisms follows from extension of existing legislation as described above for environmental releases in the context of field testing. The efficacy of this approach to regulating commercial use of genetically modified organisms is a debated issue. The wisdom of alternative regulatory approaches will become more clearly defined once the results of field trials and ecological risk analyses become available. In light of rapid technical development, however, it is not too early to begin defining a commercial use permit system. Such a system was provided in U.S. Congressional bills for the Transgenic Animals Regulatory Reform Act of 1988 and the Omnibus Biotechnology Act of 1990. Whether or not legally mandated, the effect of Congressional concern so evidenced is to spur regulatory agencies to take definitive action.

With the commercialization of genetically modified organisms rapidly becoming a technical possibility, definition of regulatory mechanisms is timely and badly needed. Indeed, the Vice President's Council on Economic Competitiveness has considered the ramifications of current and proposed regulations of biotechnology.

Regulatory structures in other countries: In the Netherlands, commercial use of genetically modified organisms is subject to a permitting requirement administered by the Ministry of the Environment under regulations adopted early in 1990 (Hodgson, 1990).

Within the EEC, a draft directive on transgenic animals is being prepared by the Agricultural Directorate (Hodgson, 1990a). An industrial group has complained about provisions that would require manufacturers to submit data not only on product quality, safety, and efficacy, but also on potential socioeconomic effects before receiving approval for marketing genetically modified organisms.

2. Patenting considerations: Commercialization of transcenic fishes will be influenced by the patentability of novel life forms and resolution of several related issues (Hallerman and Kapuscinski, 1990b). We believe it likely that some, but not all transgenic lines will be protected by patents.

Key among patent-related issues is determination of the structure of royalty obligations. Because transgenic animals are, in this sense, inventions that reproduce themselves, the obligation to pay royalties to patent holders will affect the utilization of transgenic lines by the aquaculture sector. Current law fails to adequately address the respective rights of patent holders and animal owners over the offspring of patented animals (Merges, 1988). A bill for a Transgenic Animals Patent Reform Act passed in the U.S House of Representatives in 1988, but died in committee in the Senate.

A patenting issue directly affecting the conduct of scientific studies is the scope and strength of the exemption from patent protection for research purposes (Hallerman and Kapuscinski, 1990b). Present policy is based on judicial decisions, and is not well defined.

International aspects of patenting policy: Differences exist among countries regarding what constitutes patentable material. Although several international treaties and agreements exist to some degree harmonize the procedure and substance of international patent practice (Office of Technology Assessment, 1989), no existing agreement addresses the patentability of animals.

The U.S. is the only nation which has granted a patent for a transgenic animal, in this case a mouse (Hallerman and Kapuscinski, 1990b). It seems likely that other nations will issue similar patents in the future. The Japanese patent office has announced a policy of granting patents on nonhuman animals (Office of Technology Assessment, 1989). Anticipating the 1992 union of the European Economic Community, the European Commission has proposed what amounts to a European patent law for biotechnology (Newmark, 1989). In an effort to define certain terms in current statutory language so as to minimize legal restrictions, the Commission has signaled the intent to make animals potentially patentable (Lesser, 1989).

Still, any pronouncement of the number of countries permitting patenting of transgenic animals will have to await the outcome of decisions on particular patent cases (Office of Technology Assessment, 1989). In this connection, the European patent application for the U.S.-patented transgenic mouse was rejected by the European Patent Office on the grounds that the animal was socially undesirable (Hodgson, 1990a).

THE IMPORTANCE OF PUBLIC PERCEPTION

"At this stage of its emergence, the biotechnology, industry can learn from the painful experience of the chemical industry: That a public enthralled by technical advances can become quickly hostile if they are not honestly informed about both the benefits and potentially hazardous side effects and then given adequate opportunity to participate in regulatory decision making (J.D. Hair, cited in Office of Techology Assessment, 1988, page 43)."

As part of a broader survey of U.S. public opinion an issues regarding environmental release of genetically modified organisms (Office of Technology Assessment, 1987), responses to two particular questions showed public attitudes toward trangenic fish to be divided:

- If there was no direct risk to humans and only very remote risks to the environment, would you approve or disapprove the environmental use of genetically engineered organisms designed to produce larger game fish? Fifty-three percent of adults interviewed by telephone approved, and

43% disapproved such use, a lower approval rate than for any of the other applications specifically posed.

- Suppose that a new genetically engineered organism had been developed which would significantly increase farm production with no risk to humans. Would you approve the environmental use of that organism if the risk of losing some local species of plants or fish was:

Unknown?	31%	approved.
Unknown, but very remote?	45%	approved.
1 in 100?	40%	approved.
1 in 1000?	55%	approved.
1 in 10,000?	65%	approved.
1 in 100,000?	71%	approved.
1 in 1,000,000?	74%	approved.

Case histories of environmental releases in the U.S. (Office of Technology Assessment 1988; Witt 1990, Ch. 4) revealed that public attitudes about proposed experiments was important in determining whether such experiments were eventually conducted. Further, positive public attitudes depended upon perceptions that (1) proponents of the releases were forthright regarding the potential benefits and risks of the test, and (2) regulatory authorities had considered the concerns of the local community.

Studies of European public opinion (cited by Dixon, 1990b) showed a similarly wide divergence of views. About a third of those interviewed in a 1979 study by the Commission of the European Communities thought that genetic research was worthwhile, while another third thought that such work involved unacceptable risks. Another cited study showed important variation in public opinion among countries, with 61 percent of Danish respondents finding genetic research unacceptable, as compared with 22 percent in Italy.

Efforts to regulate the development of biotechnology are apparently lacking among developing countries (Table 1). Within these societies, pressures to expand food supplies by any means may supercede any desire to protect ecosystems from the potential impacts of genetically modified organisms.

DEVELOPMENT OF TRANSGENIC FISH, POLITICAL REALITIES, AND THE ROLE OF SCIENTISTS

"We must take the initiative not only in expounding the medical, agricultural, and environmental benefits of applied bioscience, but also in discussing openly whatever possible dangers are raised bv critics (Dixon, 1990b)."

"The scientific community urgently needs to provide guidance to both investigators and regulators in evaluating planned introductions of modified organisms from an ecological perspective (National Academy of Sciences,1987)."

The uncertainties associated with the development and use of transgenic fishes and the importance of public perception will affect the handling of controversies surrounding requests for releases of transgenic fishes within the next several years. Scientists could contribute to the public policy debate (1) by generating reliable data regarding performance and possible environmental impacts of transgenic fish, and (2) by advocating responsible use of transgenic fishes. In this context, the American Fisheries Society has adopted a position statement (Kapuscinski and Hallerman, 1990b) recommending: research providing data needed for rational policy decisions, caution in uses of transgenic fishes, and consolidation of regulatory authority over genetically modified organisms. Public confidence in the results of performance tests and ecological risk analyses, in addition to the results themselves, will prove critical in determining whether transgenic fish eventually reach practical utilization.

ACKNOWLEDGMENTS

We thank Gail Keenum for reviewing a draft of this paper. Support for E.M.H. while writing this paper came from the Virginia Legislative Intitiative for Aquaculture. Support for A.R.K. came in part from the Minnesota Agricultural Experiment Station, the Minnesota Sea Grant College Program supported by the NOAA Office of Sea Grant, Department of Commerce, under Grant Nos. NOAA/NA86AA-D-SG112, Project No. A/SE-1 and NOAA/NA86AA-DOSG112, Project No. RA/5, J.Rand from the Legislative Commission on Minnesota Resources. This is article 18523 of the Minnesota Agricultural Experiment Station Scientific Journal Article Series.

REFERENCES

Ager, B.P. (1988) Regulations and the European scene. Pages 542-544 in J. Hodgson and A.M. Sugden, eds. Planned release of genetically engineered organisms. Trends in Biotechnology/Trends in Ecology and Evolution. Elsevier Publications, Cambridge.

Animal and Plant Health Inspection Service. (1987) Plant pests: Introduction of genetically engineered organisms or products: final rule. Federal Register 52: 22892-22915, 16 June.

Anonymous. (1990) Euro-release regulations. Bio/Technology 8:490.

Benfey, T.J., Dye, H. M., Solar, I. I. and Donaldson, E. M. (1989) The growth and reproductive endocrinology of adult triploid Pacific salmonids. Fish Physiology and Biochemistry 6:113-120.

Carpenter, S.R. and Kitchell, J.F. (1988) Introduction. Pages 1-8 in S.R. Carpenter, ed. Complex interections in lake communities. Springer-Verlag, New York.

Connell, J.H. (1975) Some mechanisms producing structure in natural communities: A model and evidence from field experiments. Pages 460-490 in M.L. Cody and J.M. Diamond, eds. Ecology and evolution of communities. Belknap Press, Cambridge, MA.

Cooperative State Research Service. (1990) Research proposed on transgenic fish; Publication of Environmental Assessment; Notice of opportunity for public comment. Federal Register 55: 5752-5772, 16 February.

Dixon, B. (1989) Inching toward uniformity on Euro-release. Biotechnology 7: 742.

Dixon, B. (1990a) Deliberate release regulations: Scope of U.K. proposals widened. Biotechnology 8: 100.

Dixon, B. (1990b) Commentary: Seize the dais. Biotechnology 8: 487 .

Dixon, B. (1990c) Environmental regulation: Biotech's effect on biodiversity debated. Biotechnology 8: 499.

Fox, J.L. (1989) Fish drift between agencies' guidelines. Bio/Technology 7: 865.

Fox, J.L. (1990) Federal legislation: Toward unified rules on deliberate release. Bio/Technology 8: 499.

General Accounting Office. (1987) Bio/Technology: Role of institutional biosafety committees. Document RCED-8864BR, GAO, Washington, D.C.

Goudie, C. (1988) Some triploid grass carp can be induced to spawn. Research Information Bulletin Number 88-24. National Fisheries Research Center, Gainesville, FL.

Government of Canada. (1988) Biotechnology regulations: A user's guide. Ministry of State for Science and Technology, Cat. En 40-357/1988 E, Ottawa, Canada, 47 pages.

Hallerman, E.M., and Kapuscinski, A.R. (1990a) Transgenic fish and public policy: Regulatory concerns. Fisheries 15(l): 12-20.

Hallerman, E.M., and Kapuscinski, A.R. (1990b) Transgenic fish and public policy: Patenting of transgenic fish. Fisheries 15(l): 21-24.

Hallerman, E.M., and Kapuscinski, A.R. (1990c) Ecological implications of rising transgenic fish in aquaculture. Proceedings of the International Symposium on the Effects of Introductions and Transfers of Aquatic Species on Resources and Ecosystems, Halifax, Nova Scotia, 12-13 June, 1990. In press.

Hodgson, J. (1990a) Regulatory policy: European advisory group bares its teeth. Bio/Technology 8: 185.

Hodgson, J. (1990b) Deliberate release: Dutch regulations now in force. Bio/Technology 8: 284.

Kapuscinski, A.R. and Hallerman, E.M. (1990a) Anticipating environmental impacts of transgenic fish. Fisheries 15(l): 2-11.

Kapuscinski, A.R. and Hallerman, E.M. (1990b) American Fisheries Society position statement: Transgenic fishes. Fisheries 15(4): 2-5.

Kapuscinski, A.R., and Hallerman, E.M. (1990c) Implications of introduction of transgenic fish into natural ecosystems. Canadian Journal of Fisheries and Aquatic Sciences: in press.

Lesser, W. (1989) Animal patents: A ray of hope for animal patents in Europe. Biotechnology 7: 221.

McCormick, D. (1987) Japan: Coordinated action. Bio/Technology, 5 : 1 2 7 9 .

Merges, R.P. (1988) Intellectual property in higher life forms: The patent system and controversial technology. Maryland Law Review 47:1051-1107.

Moav, B., Liu, Z., Moav, N.L., Gross, M.L., Kapuscinski, A.R., Faras, A.J., Guise, K. S. and Hackett, P.B. (1992) Expression of heterologous genes in transgenic fish. In: Transgenic Fish. C. Hew and G.L. Fletcher (eds). World Science Publishing Co., Singapore, pp. 120-141.

Morgan, D. (1990) Biotech panel wrestles with oversight role. The Scientist 4(4): 1-14, 19 February 1990.

National Academy of Sciences. (1987) Introduction of recombinant DNA-engineered organisms into the environment: Key issues. National Academy of Sciences Press, Washington, D.C.

National Biotechnology Advisory Committee. (1989) The regulation of biotechnology: A critical issue for Canadian research and industrial development. Industry, Science, and Technology Canada, Ottawa, Canada. 23 pages.

National Institutes of Health. (1984) Guidelines for research involving recombinant DNA molecules. Federal Register 49: 46266-46291, 23 November.

National Institutes of Health. (1986) Guidelines for research involving recombinant DNA molecules. Federal Register 51: 16957-16985, 7 May.

Newmark, P. (1987) Discord and harmony in Europe. Bio/Technology 5: 1281-1283.

Newmark, P. (1989) Patents: A modest proposal for European patents. Biotechnology 7: 26.

Odum, E.P. (1984) The mesocosm. BioScience 34: 558-562.

Office of Agricultural Biotechnology, U.S. Department of Agriculture. (1990) Draft: Guidelines for research with genetically modified organisms outside contained facilities. Dated June 19, 1990.

Office of Science and Technology Policy. (1984) Proposal for a coordinated framework for the regulation of biotechnology. Federal Register 49: 50856-50907, 31 December.

Office of Science and Technolocy Policy. (1985) Coordinated framework for the regulation of biotechnology; Establishment of the Biotechnolgy Science Coordinating Committee. Federal Register 50: 47174-47195, 14 November.

Office of Science and Technology Policy. (1986) Coordinated framework for the regulation of biotechnology. Federal Register 51: 23301-23350, 26 June.

Office of Technology Assessment. (1987) New developments in biotechnology, 2: Public perceptions of biotechnology. OTA-BP-BA-45. U.S. Government Printing Office, Washington D.C.

Office of Technology, Assessment. (1988) New developments in biotechnology, 3: Field-testing engineered organisms; Genetic and Ecological Issues. OTA-BA-350, U.S. Government Printing Office, Washington, D.C.

Office of Technology Assessment. (1989) New developments in biotechnology, 5: Patenting life. OTA-BA-370, U.S. Government Printing Office, Washington, D.C.

Organisation for Economic Cooperation and Development. (1986) Recombinant DNA safety considerations. OECD, Paris.

Regal, P.J. (1986) Natural history in the era of genetic engineering. J.F. Bell Museum of Natural History Imprint 3(4):1-4. Bell Museum of Natural History, Minneapolis, MN.

Thorgaard, G.H., and Allen, S.K. Jr. (1987) Chromosome manipulation and markers in fishery management. Pages 319-331 in N. Ryman and F. Utter, eds. Population genetics and fishery management. University of Washington Press, Seattle.

Tiedje, J.M., Colwell, R.K., Grossman, Y.L., Hodson, R.E., Lenski, R.E., Mack, R.N. and Regal, P.J. 1989. The planned introduction of genetically engineered organisms: Ecological considerations and recommendations. Ecology 70: 298-315.

Witt, (S.C. 1990) Biotechnology: Microbes and the environment. Center for Science Information, San Francisco, CA.

CHAPTER 13

Containment of Genetically Altered Fish with Emphasis on Salmonids

Robert H. Devlin and Edward M. Donaldson

ABSTRACT

Permanent genetic alterations to fish stocks by selective breeding and transgenesis has the ability to make major improvements in performance, and the latter technology has the capability to induce novel phenotypes. Protection of wild populations from genetically altered fish requires containment of the reproductive potential of cultured stocks and can be achieved by developing effective physical and biological controls. Although physical containment measures can be effective, they are subject to disruption by a variety of natural and human activities (ie. weather, human error), and as a consequence, biological containment measures should also be implemented. Effective biological containment can be best achieved by sterilization that can directly eliminate the potential for reproductive interaction between wild and cultured stocks. A variety of sterilization procedures have now been developed, and, of those applicable to commercial aquaculture operations, induction of triploid females and sterilization with high doses of androgens have the most promise for application in the immediate future. Induction of sterility through the use of transgenic fish containing engineered genes capable of blocking normal reproductive development also has considerable potential for allowing effective reproductive containment of genetically altered fish.

INTRODUCTION

The implementation of biotechnology into world aquaculture production systems has now begun, and it is anticipated that such efforts will have an increasing impact on fish production in the near future. Although major advances can also be foreseen in other aspects of aquaculture technology, the application of biotechnology to aquaculture is the focus of a large international research effort and has recently attracted significant public attention. Biotechnological improvement involves the production of modified organisms that differ in some way from natural populations. These modifications are designed to enhance one or more desirable characteristics of the cultivated fish and can be categorized into either permanent changes (genetic alterations) or transient changes (treatment with bio-active compounds). While both of these groups require similar consideration for implementation, genetically altered fish raise the most concern due to the potential for long-term, sustainable changes. It is imperative that coupled with biotechnology research and development programs there be an equally strong and parallel effort designed to safely utilize this new technology such that human health and the natural environment are not compromised. This will require simultaneous development of effective regulatory guidelines for control and pertinent scientific research programs that address potential problems and develop viable solutions.

The protection and genetic conservation of natural fish fauna is an important issue from a variety of social and biological perspectives. Wild fish stocks are considered a valuable renewable resource for recreational and commercial fisheries, both of which are controlled by resource managers. These wild stocks also provide a valuable source of genetic material for the development of new aquaculture strains. Effective guidelines for the implementation of modified fish must satisfy the concerns of both the public and the regulatory agencies and, at the same time, not be so restrictive that important biological enhancement technologies are unnecessarily restrained. Thus, when deciding upon implementation guidelines and regulations, the relative merit of each genetic modification must be weighed against the potential impact on fish populations. Careful evaluation of critical parameters prior to implementation of new technologies can allow for their effective use with minimal impact. To a large extent, such evaluations have not been performed adequately in the past or have been considered to be of secondary importance (e.g., to health requirements) when, for example, new industrial developments or fish transplantations are considered.

The objectives of this paper are to briefly discuss concerns associated with the implementation of genetically altered fish species, and to outline the current status and future prospects of containment methodologies that have been developed for fish (with emphasis on salmonids) which could effectively contain such fish populations.

GENETICALLY ALTERED ORGANISMS REQUIRING CONTAINMENT

Several types of genetically altered organisms requiring effective containment can be envisaged. Each possesses different characteristics that pose distinct threats and which may require different measures to regulate and control their implementation.

Alterations to the genetic constitution of fish can occur at the level of individuals, populations, or species, and may be divided into two broad groups. The first group, which is by far the most common, includes those alterations that change the arrangement or organization of native genes within individuals or groups of individuals. The second group requiring consideration is that where new genetic information has been introduced into the species (for example, see Fletcher et al., 1988, 1992 (Chapter 11, this monograph)). This latter group falls into 2 categories: the introduction of additional copies of genes derived from the host species, and the introduction of foreign DNA sequences that do not occur naturally in that species.

Alterations to the genetic structure of fish populations (see Table 1) routinely occur through natural selection, driven by processes such as environmental changes to habitat (drought, fluctuations in temperature or other physical or chemical parameters), changes to feed resources, or predator-prey interactions. Many activities undertaken by man may also influence the genetics of fish populations; for example, industrial developments that alter survival or reproductive capacity, recreational and commercial fishing, and programs that involve the hatchery rearing of fish for enhancement of wild stocks or for utilization within the aquaculture industry. Changes to the genetic structure of fish stocks can also be purposefully brought about through selective breeding programs which are designed to improve particular production characteristics over a period of generations. The population genetic mechanisms that may operate in such alterations include inbreeding, genetic drift arising in small populations, and either natural or artificial selection for alternative genotypes.

Fish populations are locally differentiated with respect to morphological characteristics and life history traits, and this differentiation is believed to arise from adaptive responses to regional differences in environmental conditions. Indeed, studies have shown that, for several fitness parameters, local, native stocks invariably perform better than do introduced strains (Bams, 1976; Reisenbirchler and McIntyre, 1977; Fraser, 1981; MacLean et al., 1981; Chilcote et al., 1986). If such traits are largely determined by genetic parameters, then alterations in fitness might be expected when new genetic information is introduced into a population from a stock that is not adapted to the same conditions (see Ferguson, 1990). Hybrids may in fact not resemble either of the parental fish (Ferguson et al., 1988).

It can also be argued that positive effects on population fitness may arise, in the initial generations, after the introduction of new genetic information, either from hybrid vigour or by providing increased genetic variation (in the form of favourable alleles that were not previously available to the population) that could undergo selection to further improve the stock. However, because significant levels of straying exist in natural salmonid populations (Quinn, 1984), genetic variation arising in one stock is eventually capable of reaching all populations, to be either eliminated or incorporated into its genetic structure by local selective forces. This process occurs continuously, and therefore most genotypes have probably had the opportunity to be selected throughout much of the geographic range of the species. Consequently, it might be expected that the majority of infusions of non-native genes would disrupt favourable population

Table 1. Examples of genetic alterations of fish populations

Type of genetic alterations	Concerns for implementation and possible consequences of escaped fish
Selective breeding	- altered phenotype and genotype relative to wild stock that changes interaction and competition within the community
Genetic drift	- altered gene frequencies in wild population arising from interactions with improved stocks derived from small populations
Inbreeding and outbreeding	- inbreeding and outbreeding depression of wild stocks by extensive hybridization with non-native strains
Polyploidy	- degree of sterility, ability to interfere with spawning, competition for resources without contributing to subsequent generations
Hybridization	- ability to transfer genetic information between species
Mutation	- altered phenotype; large increase in mutant gene frequency, probably associated with inbreeding
Transgenic	- characteristics of the phenotype could include some of the above concerns - characteristics of the genetype - function of the gene - type of gene introduced - how regulation has been altered - source of coding DNA - source of regulatory DNA - potential for causing pleiotropic effects - potential for interacting with remainder of genome - stability of construct (rearrangements) - ability of construct to transpose within or between genomes

genetic structures and have a negative impact. This appears to be the case (see Gausen and Moen, 1991) for a number of indigenous salmonid populations that have interbred with non-native strains (Bams, 1976; Nickelson et al., 1986; Taggart and Ferguson, 1986; Altukhov, 1981; Altukhov and Salmenkova, 1987). Such negative or less fit genotypes may not always be removed from a population, particularly if the number of introduced individuals is large relative to the indigenous population or there are frequent repeated introductions. A study by Philipp (1990) clearly demonstrated that local largemouth bass outperform introduced stocks with respect to survival, growth, and reproductive success. This work also showed that subsequent generations, composed largely of hybrids between indigenous and introduced strains, displayed significantly poorer performance traits. It is not known how long such a condition of stock degeneration will persist, or if the local environmental conditions will be able to reselect the appropriate favourable genotypes.

Altered genotypes with new or rearranged genetic information can be produced by natural or artificial hybridization, mutation, or gene transfer techniques. These alterations have neither more nor less ability to disrupt ecosystems than the genetic changes outlined above, but can be subjected to greater scrutiny due to their method of production. It is apparent that more objections would be raised to a chinook salmon that had a single sockeye salmon gene inserted into its genome (transgenic) by recombinant DNA methods than to a chinook hybrid containing an entire sockeye genome produced by natural or artificial interspecific hybridization. It is extremely important that regulators distinguish between the characteristics of a new stock of fish and the biotechnological process that was used to produce it. It is the phenotypic and genotypic characters of a fish strain that should be scientifically evaluated for suitability for introduction into the environment, whereas consideration of the process utilized to make such changes is an ethical question requiring consideration by society at large.

Transgenics, however, also require special attention because they have the potential to contain new genetic information (see Table 1) that would not be easily acquired by the species during the course of evolution. This new information may present long-term concerns (such as gene promoters with novel or multiple regulatory properties, or viral coat protein genes inserted to provide disease resistance), and should therefore be carefully evaluated with respect to the potential for influencing phenotype in the future. It should also be noted that transgenic fish not only contain introduced genetic information, they have also undergone a severe bottleneck in their production, since each stock is derived from only a single fish. Such strains, therefore, will lack the natural variation present in wild populations unless deliberate efforts are made to reintroduce it through extensive outcrossing.

The genetic concerns relating to fish introductions are dependent upon such fish being able to reproductively interact with indigenous populations. In the case of escaped farmed Atlantic salmon (NASCO, 1990), several studies have examined the migration of these fish into spawning streams in Norway (Hansen et al., 1987; Furevik et al., 1990; Egidius et al., 1991; Gausen and Moen, 1991), and although their reproductive success is unknown, they are believed to interbreed with native populations. The ability of farmed salmonids to return to streams to spawn is highly dependent on the age and physiological state of the fish, and on the location and type of water they are released into. Escaped farmed fish tend to return to rivers in the vicinity of their release, although straying rates and distances increase the longer the fish have been penned in seawater. Smolts that are transferred to sea-cage sites without experiencing the smolt migration were found to stray more and were not found to return to their natal river system as well as did fish allowed to migrate to sea from the river's mouth (Hansen et al., 1987). Farmed fish appear to enter spawning streams later than wild stocks, they do not migrate as far upstream or bypass obstacles as well as wild fish, nor do they possess the typical phenotypic traits associated with, and presumably advantageous for survival in, different spawning streams (Gausen and Moen, 1991). Such reductions in fitness of farmed fish relative to wild stocks may provide a certain degree of reproductive isolation. Skaala et al. (1990 and 1991) have examined the reproductive success of farmed brown trout (*Salmo trutta*) in Norway, and have found that farmed trout were capable of breeding with themselves and with wild trout, although with lower success than their wild counterparts.

ECOLOGICAL CONCERNS RELATING TO PHENOTYPE AND GENOTYPE

The ecological concerns relating to introductions of genetically altered organisms have received careful consideration in several recent reviews (Tiedje et al. 1989; Ferguson 1990; Kapuscinski and Hallerman 1990; Hallerman and Kapuscinski, 1992, chapter 12, this monograph). The problem is complex, encompassing a wide range of biological processes and fields of study, including, for example, genetics, ecology, and genotype-environment interactions. The effect of genetically altered fish on the ecosystem will occur at several levels, from direct effects on other wild conspecifics in the population to direct or indirect effects on other species that interact within the community. Additional mechanisms by which introduced fish can interfere with natural populations have been discussed (Moyle, 1986; Tiedje et al. 1989; Kapuscinski and Hallerman 1990) and several examples relating to salmonid biology are shown in Table 2. The degree of interaction will depend on the fitness of a particular phenotype, the stability of the genetic alteration, the number of interacting individuals, and the environment into which the introduction has occurred. Both Tiedje et al. (1989) and Kapuscinski and Hallerman (1990) suggest that, with respect to the introduction of modified organisms, phenotype is a critical factor influencing potential impacts. Phenotypic changes can arise from genetic and/or environmental factors that influence the development or physiology of an organism and it is this outward character that ultimately interacts with the bio-environment.

Although our ability to accurately predict phenotype from genotype is limited, careful evaluation of the genotype of modified organisms is also warranted. It is apparent that genotypic changes to organisms can mediate phenotypic changes (with either single or pleiotropic effects), and that such phenotypic modifications may result in some ecological impact. It is equally true that many genotypic alterations will appear functionally benign and will have no detectable affect on the phenotype under particular conditions. Such genetic changes might therefore be considered as merely "genetic pollution" and inconsequential from an ecological perspective. However, it is very likely that each genotypic alteration will result in a phenotypic change that is dependent upon both the genetic background and the environment into which the fish is introduced. Such genotypic alterations may be beneficial in one environment and deleterious in another (for example, genes conferring freezing resistance may provide an increase in fitness in sub-zero conditions, but be an unnecessary burden in warmer conditions). Similarly, particular genetic alterations may provide the desired phenotypic change in one genetic background and not in another (for example, growth hormone genes may have a marked effect in wild genetic backgrounds, but have a reduced effect in genotypes that have been selectively bred for improved growth performance).

Certain genetic changes may require long periods of time to affect phenotypes, for example, at the extreme, a DNA sequence that does not code for any functional product but may interact physically with the remainder of the genome (through recombination or transposition) to promote genetic alterations that do have a phenotypic effect. For example, in transgenic studies, the utilization of DNA sequences derived from the species being altered has the advantage of not adding "new" information to the genome, but the disadvantage of being subject, in the long term, to recombination or gene conversion with homologous sequences elsewhere in the genome. The ability of scientists to accurately predict or assess the true impact of a particular phenotype is limited, and the task would be considerably more difficult if provided only with information relating to the genotypic alteration and its effects under a limited range of environmental conditions. Clearly therefore, careful consideration of the potential short and long-term effects of both genotype and phenotype are required when modified fish stocks are considered for implementation into wild or semi-wild environments.

The persistence in natural populations of genetic alterations arising from escaped farmed fish depends to large extent on the number of individuals that successfully reproduce with local natural populations or with themselves. If interbreeding with native fish is not possible (due to physical or biological limitations) then the extent of the impact will be limited to that caused by competition within the ecosystem. This effect should not be underestimated since many examples exist where successful introductions of fish species have displaced local varieties and established feral populations (Welcomme, 1981 and 1984).

Since we are unable to effectively predict the ecological outcome of genetically altered fish in natural environments, a conservative approach should initially be adopted to prevent such introductions from occurring. Thus, effective containment measures must be co-implemented when genetically altered fish are utilized in situations where their escape has the potential to release them into environments where they can no longer be controlled.

Table 2. Examples of ecological impacts arising from specific phenotypic changes.

Phenotypic change	Potential ecological impact
Increased food requirement	- Depletion of food supply through increased competition - Changes in the trophic web
Increased reproductive capability	- Disruption of natural mating systems and competition for spawning sites
Decreased reproductive capability	- Failure (of salmonids) to provide nutrients to natal streams
Tolerance of environmental factors	- Creation of new feral populations which may compete with native species
Increased waste production	- Positive or negative effects on organic loading in the environment

METHODS OF CONTAINMENT

The level of containment required to allow safe implementation of genetically modified organisms depends on the degree to which they threaten the natural environment. Failure to effectively contain genetically altered fish could result in transient effects on the ecosystem resulting from direct competition of escaped individuals with other members of the community, or more long-term effects resulting from genetic introgression of the modifications into the natural gene pool of conspecifics. Both of these possible interactions require regulation, and could be effectively controlled by physical separation of the genetically altered individuals from the remainder of the population. Unfortunately, physical containment of fish stocks cannot be accomplished with complete certainty except in specially constructed quarantine facilities. Most fish rearing facilities are subject to environmental insults (for example, storms in the case of sea-based farms, or flooding of land-based facilities) that are capable of inflicting sufficient damage so some individuals may escape. When such events occur, the escaped fish may have a temporary effect on the local environment, and if reproductively competent, may have long-term effects as well. Consequently, the abolition of reproductive capability is also required and this can best be accomplished through biological containment methods. The most certain of these methods is sterilization, and several procedures have now been developed that are being, or could be used, on a commercial scale.

1. Physical Containment

The first line of defense against the escape of modified organisms (or their viable gametes) into the natural environment is to establish both effective physical structures as well as proper management procedures that ensure control over the movement of fish and maintenance of equipment. Many considerations for designing effective containment measures are common to quarantine procedures used to prevent the introduction of diseases (Rosenthal, 1986).

The type of physical structure required for effective containment depends on the location and environment that the fish are introduced into. Clearly, little containment is required for fish that cannot survive outside of man-made facilities, for example tropical species being cultured in heated water in more temperate regions. However, it is more usual that local species, or introduced species that grow well in the region, will be cultured, and thus the consequences of escape are more pronounced. In such cases it may be desirable to establish "aquaculture free" and "aquaculture intensive" zones to facilitate better control and minimize the large-scale impact of any fish escapes.

In the case of salmon, most strains are able to grow in the natural environment in which they are farmed. Initial growth takes place in fresh water at land-based facilities, where eggs are incubated indoors at hatcheries, moved to indoor or outdoor tanks or troughs as fry, and grown to the smolt stage in larger tanks prior to transfer to seawater. While the fresh water phase usually lasts from 6 - 12 months, the sea-water phase can last as long as two years and is usually conducted in pens consisting of nets suspended from floating structures.

Containment in land-based facilities can readily be accomplished by screening in and out-flow water ways to prevent direct escape of fish from ponds or tanks. All screening should consist of at least duplicate, independent systems and water should ideally drain to a ground filter system that provides little opportunity for fish survival if escapement does occur. Inclusion of a chlorination system in the outflow pipes (treatment for ten minutes at greater than 5 ppm free residual chlorine) can ensure that any escaped fish or gametes will be destroyed before reaching the natural environment (Gary Hoskins, personal communication). Such a quarantine system was developed to allow the introduction of disease-free Atlantic salmon to Canada's Pacific coast. Adequate protection from predators (for example over-pond bird nets, and above and underground fencing) that could transfer living individuals to natural ecosystems must also be ensured. In outdoor facilities, the installation must offer protection against natural environmental challenges (eg. flooding) that could obviate otherwise sufficient levels of containment. Although this is probably best accomplished by choosing an appropriate site for the rearing facility, additional measures such as meteorological monitoring, dikes, drainage ditches, and backup holding and catch-basin facilities should also be incorporated into development plans. Such measures have been included in the design of containment facilities (Federal Register, 1990) for a land-based population of transgenic carp containing a rainbow trout growth hormone gene (Zhang et al. 1990).

Sea-pen facilities pose a greater problem for containment than land-based sites. Accidental releases from land-based sites can be checked with backup containment systems, and it may be possible to locate and eliminate escaped individuals. Clearly, however, individuals that escape from sea (or large lake) facilities would be very difficult, if not impossible, to recover. For this reason, sea-pen releases of genetically altered organisms should be performed with the utmost of security and containment measures in place. The physical structure of net pens varies considerably throughout the industry. Most fish escapes from sea-pen facilities arise from environmental damage from storms and ice, from accidental damage by boats and harvesting equipment, accidents during fish transport, from predators tearing holes in the primary nets in search of food, or from human error, vandalism, or the deliberate release of "unprofitable fish" to claim on insurance. Although some of these problems can readily be solved by effective management practices, it is damage by unpredictable environmental actions which are of most concern. Net pens are usually rated for their ability to withstand wave action of various magnitudes. Commercial facilities will often consider these specifications and associated costs relative to the probabilities of encountering such conditions over a fixed investment or insurance period. In other words, a farm site is usually not designed to be as safe as is necessary to contain fish, but rather is capitalized to the extent that is required to ensure financial returns. For the implementation of genetically altered fish, such monetary considerations must play a secondary role, and biological concerns must be met with as close to absolute certainty as is possible. Significant improvements in secure net pen structures have recently been developed, and a variety of design types and anchoring mechanisms are in use. Open-ocean net-pen structures are now in operation and, in contrast to their inshore counterparts, they can withstand very heavy seas by moving with the water rather than trying to withstand its force (for example, see Loverich, 1991). Some structures are completely contained with netting on all sides (Willinsky et al., 1991). If such structures were to break free, and were negatively buoyant, escapement of fish would be less likely than from pens lacking netting on their top side.

Containment at any facility is subject to disruption caused by human error, lack of maintenance, theft, and vandalism. Some of these problems can be overcome with effective security measures, however mistakes made by even trained personnel are extremely common. Safeguards to minimize this problem include adequate training programs and regular evaluation of performance. Accurate record keeping of the location and number of contained fish is essential, and it must be stressed that the reporting of any releases is required, and, should not be concealed as might be the natural tendency. Compliance with regulations could be ensured by periodic external inspection and certification by industry and government agencies of the performance and integrity of any aquaculture facility containing genetically altered fish.

2. Biological Containment

The inherent and inevitable potential for failure which characterizes some physical containment systems (eg. net pens) requires the co-implementation of additional backup protection. The primary concern related to escapes of genetically altered fish is the potential to introduce new genetic material or alter the existing genetic structure of natural populations. Clearly, the most secure method for preventing reproductive interaction with wild individuals is to induce complete sterility in the stock for which containment is required. A variety of methods have been described for sterilizing salmonids and other fish. The utility of any particular method of biological containment in aquaculture operations depends first on the efficacy of the treatment, second on the ease of the technique required for induction of the sterile phenotype, and third on the economic impact on the production system. To be a truly useful sterilization method for aquaculture, a technique must be completely effective, simple and cheap to implement, and not adversely affect the performance of the fish. Systems that meet some of these requirements are now being developed for salmonid culture, and these and several other methods that are available or are in the experimental phase are shown in Table 3.

Table 3. Methodologies for producing sterile salmonids.

Method	**Suppression Implementation of maturation**	**Efficacy of sterile induction**	
Triploidy	Female - Yes Male - No	High	Simple, large scale possible
Androgen Treatment	Yes	High	Simple, large scale possible
Surgery	Yes	Complete	Labour intensive
Chemosterilization	?	Incomplete	Experimental
Autoimmune Sterility	No	Incomplete	Experimental
Sterile Hybrids	Possible	Variable, depending on species	Simple
Irradiation	No	Complete	Simple, causes developmental abnormalities
Transgenesis	?	?	Experimental

Methodologies for the sterilization of salmonids were the topic of a recent symposia (Pepper, 1991) and reviews discussing sterilization in salmonids include Donaldson and Hunter (1982a), Hunter and Donaldson (1983), Thorgaard (1983 and 1986), Utter et al. (1983), Yamazaki (1983), Donaldson (1986), Bye and Lincoln (1986), Purdom (1986), Donaldson and Benfey (1987), Secombes et al. (1987), Chourrout (1987), Piferrer and Donaldson (1988), Benfey and Donaldson (1988), Johnstone (1989), Dunham (1990) and Donaldson et al. (1991a, b).

The application of sterilization methodologies to aquaculture can provide several advantages independent of providing genetic containment in the case of escape. In the case of salmonids, the physiological changes associated with reproductive maturation are blocked by certain methods of gonadal sterility induction (see below) and the life span can be approximately doubled prior to the fish undergoing a natural process of senescence (Solar and Donaldson, unpublished). This can be a considerable advantage to aquaculturists who wish to produce a larger product and market their fish after the normal period of sexual maturation. Such strategies can allow the farmer to produce a higher value product (larger fish tend to command a better price) and provides the flexibility to market during the "off season" relative to the wild fishery. In addition, the sterilization of production fish could allow protection of proprietary genetic strains with improved production characteristics that are now being produced either by transgenesis and/or by selective breeding. Broodstock populations of such genetically desirable strains could be maintained under controlled conditions, whereas fish derived from these stocks to be sold to the aquaculture industry could be sterilized to prevent the unauthorized propagation of the genetic material.

A variety of sterilization methods have been or are being developed, and their potential for application to aquaculture and genetic containment are described below.

A. Chromosome Set Manipulation

Chromosome set manipulation refers to the production of polyploid individuals with additional complete sets of chromosomes (Thorgaard, 1986). Although in other fish and non-fish species this condition does not necessarily cause sterility, in salmonids, both male and female triploids are effectively infertile (Benfey and Donaldson, 1988). Triplody in females appears to arrest, for the most part, the process of oocyte development, since vitellogenic oocytes are rarely observed in triploid ovaries. The absence of maturing oocytes in triploid females also appears to prevent sexual maturation: these fish presumably do not have the gonadal capacity required (i.e. lacking granulosa cells in the follicular epithelium to synthesize sufficient estrogen) to elicit development of secondary sexual characteristics associated with reproductive development and maturity. In males, gonadal development does not seem to be seriously impeded and fully mature triploid salmon producing milt are normally obtained in this sex. Milt from triploid males, however, does not produce viable offspring when used to fertilize regular female ova. The exact cause of sterility in salmonids is not known, but the large number of chromosomes possessed by these species (2n approximately = 60) makes the probability of obtaining a euploid gamete extremely rare (less than 10^{-9}, assuming random segregation of trivalents). Presumably, euploidy, or close to it, is required for normal oocyte development, whereas spermatogenesis appears to be more tolerant to the aneuploid condition. Most gametes produced by triploids males have between 1 and 2 chromosome sets (Benfey et al., 1986) and when joined with a normal 1n gamete would produce inviable aneuploid individuals with chromosome numbers between 2n and 3n. The few eggs that are obtained from triploid females are also aneuploid and do not produce viable offspring (Johnstone et al., 1991). Each triploid strain should be evaluated carefully however, since some viable offspring have been obtained from allotriploid hybrid males (Ueda et al., 1991).

Triploidy can be induced by causing either the male or female gamete to retain two sets of chromosomes (diploid gamete) rather than the usual single set. When such a diploid gamete is fused with a regular haploid gamete at fertilization, a triploid zygote will result. Such zygotes are chromosomally balanced, and, except for the sex-determining region on the Y-chromosome, have the same number of different chromosomes relative to diploids.

In females, the production of diploid gametes is usually achieved by causing the retention of the second polar body. During the second phase of meiosis, an equational division produces a haploid oocyte nucleus and a haploid "mininucleus", or polar body, that is normally shed from the oocyte after fertilization. Treatment of fertilized eggs with a physical or chemical shock for a short duration at the beginning of the phase between fertilization and the first cleavage can result in the retention of the second polar body, and this appears to fuse with both the male and female pronuclei to produce a triploid zygote nucleus. The type of shock treatments usually involve the application of either elevated temperature or pressure, however treatments with polyethylene glycol to cause dispermy (Ueda et al., 1986), nitrous oxide (Johnstone et al., 1989), and electric current plus temperature shock (Teskeredzic et al., 1991) have been used. In general, pressure shocking has been adopted as the method of choice because of the uniformity of the treatment which can be applied. Large-scale pressure shocking devices have been developed to facilitate application of this methodology to commercial situations (Benfey et al. 1988; Johnstone et al. 1991; Jungalwalla 1991) However, temperature treatment has the advantage that it doesn't require any specialized equipment, and, although the treatments are usually more variable, they can be more simply conducted on a large scale and at a variety of commercial sites.

The efficacy of triploid induction is highly dependent on the exact conditions employed. An extensive comparison of the effect of different times, temperatures and pressures on the induction of triploidy in *Oncorhynchus mykiss* have been reported by Guoxiong et al. (1989). In general, more severe treatments are more efficient at inducing triploidy, but this result is compromised by a reduction in embryonic viability. For example, with 26°C temperature shocks, a 20 minute treatment gave 48.8% survival and 100% triploidy, whereas a 10 minute treatment gave 73.1% survival but only 69.5% triploids. Similarly, the only group to yield 100% triploids with a pressure shock of 7.6 kPa for 6 minutes had very poor survival (27%). For genetic containment purposes, it will be necessary to apply conservative treatments that can ensure 100% triploidy, perhaps at the expense of embryonic viability. In many hatchery facilities, eggs are not a limiting factor, and trading off embryonic survival for completely sterile populations would be feasible. Johnstone et al. (1991) have established proprietary conditions for *Salmo salar* that allow for 100% triploid induction with 90% survival.

Diploid gametes have also been produced from males through chromosome set manipulation techniques that allow the production of tetraploids. These polyploids have been generated in salmonids (Thorgaard et al., 1981; Chourrout, 1982 and 1984; Chourrout et al. 1986; Myers et al., 1986), tilapia (Myers et al., 1986; Don and Avtalion, 1988), carp (Reddy et al., 1990), and catfish (Bidwell et al., 1985) and are usually produced by applying a late pressure or temperature shock just preceeding the first mitotic division. This can prevent karyo- and cytokinesis and double the number of chromosomes present in the embryonic nuclei. Although these animals are difficult to produce and maintain, they should generate diploid gametes that can be fused with regular haploid gametes to produce triploid zygotes. This has been experimentally demonstrated for rainbow trout tetraploid males (Chourrout et al., 1986) and carp (Wu et al., 1988), however some difficulty may exist for the larger sperm head to penetrate the micropyle during fertilization. The production of triploids by mating tetraploid females with diploid males has also been reported (Chourrout and Nakayama, 1987; Wu et al., 1988).

From the perspective of genetic containment, both male and female triploids are gametically sterile. Unfortunately, since males mature sexually, they could return to spawning grounds and compete with normal fertile males for mates and spawning territories. Female triploids would be preferred for containment purposes because only a small proportion undergo sexual maturation (about 1 per 1000 show signs of partial maturation; see Johnstone et al., 1991) and any individuals that escape to the wild would not migrate to natal streams to spawn. If the very few that do partially mature were to return to spawn, they are also gametically sterile and cannot produce viable offspring.

All-female populations of triploid salmonids can be produced either by direct feminization with estrogen treatments, or by utilizing genetically manipulated all-female strains. An alternative to the direct treatment of food fish with hormones to control their phenotype for the implementation of all-female populations of triploid salmon involves the production of monosex populations of genetically female, phenotypically male broodstock. These broodstock individuals produce only female sperm (only X-bearing) and when crossed to regular females will produce all female progeny. This technology has been extensively used to produce monosex female populations of chinook salmon for the British Columbia aquaculture industry (Hunter et al., 1983), and Johnstone et al. (1991) have produced all-female populations of triploid Atlantic salmon by pressure shock for the Scottish industry. All-female triploid populations could also be generated for production purposes by mating existing diploid genetically female phenotypic males with tetraploid female broodstock. Alternatively, all-female populations of tetraploids could be produced and maintained by masculinization of tetraploid stocks (by androgen treatment) coupled with selection of the genetically female males by test crossing, hermaphroditism, or the use of a Y-chromosomal DNA probe (Johnstone et al. 1978; Okada et al. 1979; Hunter et al. 1983; Johnstone and Youngson, 1984; Devlin et al. 1991). From this stock, both the male and female tetraploids could be used to produce all-female triploids when crossed with genetically female diploids. In Figure 1. we present a potential scheme for the implementation of genetically altered fish that are contained through chromosome set manipulation methodologies.

Of critical importance to the aquaculture industry is the performance of triploid fish relative to diploid counterparts. Extensive experimental and field trials have shown that, overall, the growth performance and viability of triploids are comparable to non-maturing diploids (Solar et al. 1984; Benfey et al. 1989; Johnstone et al., 1991; Jungalwalla, 1991; Sutterlin and Collier, 1991; Boulanger, 1991). Diploid fish, and triploid males, undergo sexual maturation and experience a growth spurt due to the presence of anabolic sex-steroids and are therefore preferred for culture situations requiring a rapid growth phase within one year at sea. The use of monosex male triploid populations could allow combination of genetic containment and this maturational growth spurt. Monomale populations could be produced either by direct treatment with androgen or by the production of "supermale" (YY) fish (Hunter et al. 1982a; Parsons and Thorgaard, 1985; Chevassus et al., 1988).

Although non-maturing fish (i.e. triploid females) do not grow as fast as their maturing counterparts, they do attain a larger final weight and maintain their silver-bright appearance and marketability throughout the year. Boulanger (1991) has noted that improved flesh quality is associated with triploidy in brook trout and that any additional expense derived from increased egg costs are more than offset by improvements in feed conversion efficiencies and marketable carcass quality and weight. The performance of triploids also appears to be amenable to stimulation with exogenous growth hormone (McLean et al., 1991), suggesting that triploids should also be effective for the genetic containment of transgenic strains with additional growth hormone genes without compromising their performance characteristics.

B. Hormonal Ablation of Maturation

Treatment of salmonids with androgen can be used either to control sexual phenotype or to induce sterility. Low doses of androgen applied by immersion during early development are effective at converting females into phenotypic males (Baker et al., 1988; Piferrer and Donaldson, 1991) This procedure is commonly employed for the development of monosex broodstock populations since the phenotypic males that are genetically female will produce all-female progeny when crossed to regular females (Hunter et al. 1983). Higher doses of androgen, usually applied as a food additive alone or in combination with early immersion treatments can result in sterilization of both genetic males and genetic females (Simpson, 1975 -1976; Johnstone et al., 1978; Hunter et al., 1982; Johnstone, 1989). Oral treatments are usually applied from first feeding until a time when the normal differentiation of the gonad is well under way; in salmonids this is approximately 2 to 3 months. Treatment with androgen by immersion of alevins alone has not yet been successful in producing all-sterile populations of salmonids. The efficacy of androgen sterilization is improved when oral application is combined with immersion treatment: for example treatment of coho salmon eyed eggs

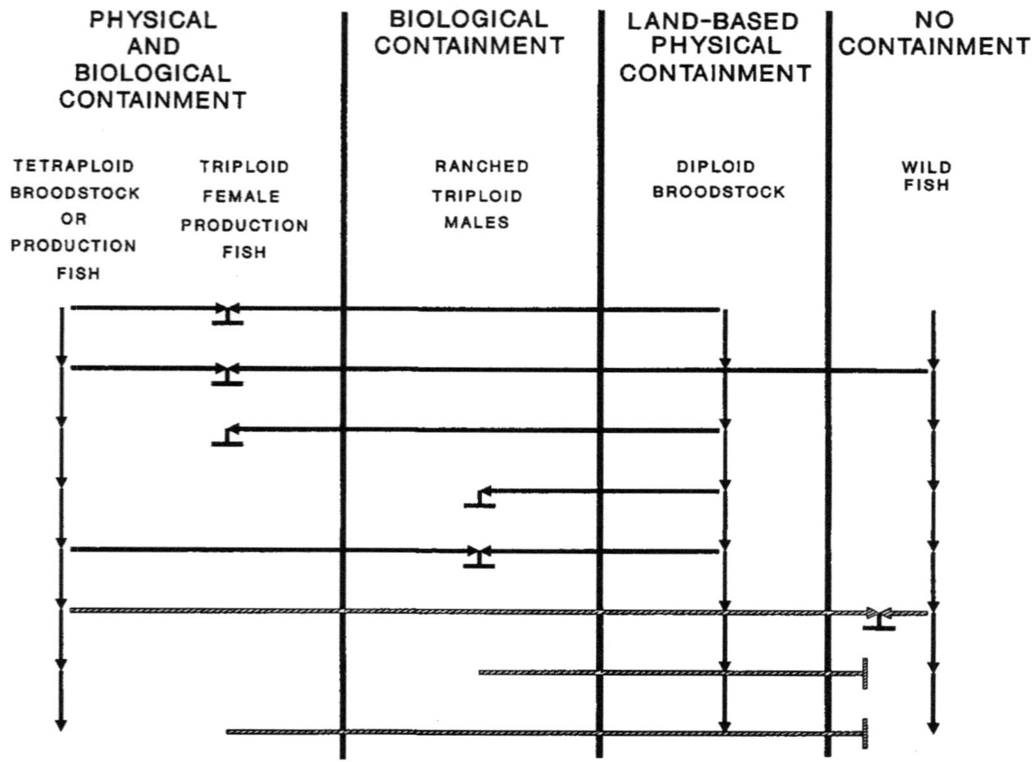

Fig. 1. Containment of genetically altered fish using chromosome set manipulation. Vertical arrows indicate reproductive success, solid horizontal arrows indicate crosses used for production of strains, and hatched horizontal lines indicate escapes or introductions.

with 400 ug/L of 17-alpha-methyltestosterone at the eyed egg stage followed by dietary administration with 20 mg/kg food for three months resulted in 100% of the individuals possessing undifferentiated gonads (Goetz et al. 1979). The dosage required to produce a sterile phenotype is dependent on the fish species, the type of androgen employed and the duration of treatment (see Table 4). Since Piferrer and Donaldson (1991) have recently shown that non-aromatizable androgens are extremely effective at inducing masculinization of genetic females, it will be important to evaluate the potency of these compounds in sterilization studies relative to the natural aromatizable analogs. These androgens also have the advantage of not being capable of conversion to estrogen and thus causing paradoxical feminization (Solar et al., 1984; Piferrer and Donaldson, 1991). Other androgens such as stilbestrol (Eckstein and Spira, 1965) and 4-chlorotesterone (Hirose and Hibiya, 1968) have also been shown to be effective in sterilizing fish.

Sterilization by androgen treatment shows great promise for application to commercial aquaculture. In addition to the benefits to be gained through suppression of maturation, this method has already been shown to prevent the reproductive interaction of sterilized fish (released into the wild) with natural indigenous stocks (Solar et al., 1986; Baker et al., 1989). In these studies, no sterilized salmon returned to the hatchery of origin, however a small proportion of incompletely sterilized fish did return. Sterilized salmon continued to grow in the natural environment and could be recovered in the wild fishery for up to five years after release.

Although the procedures required to implement androgen-sterilization technology are straightforward, the achievement of 100% sterile, large scale populations of salmon remains somewhat inconsistent. Until this goal has been achieved, the method must be considered experimental, and further research must be undertaken to demonstrate large-scale applicability alone or with other sterilization methodologies (for example combining triploidy and androgen treatments). In addition, some reluctance to use hormonal sterilization methodologies exists due to the negative social stigma associated with the use of "hormones" in the production of food animals. Such concerns appear to be scientifically unfounded since any residue of applied hormone has been completely eliminated from the fish's body within three weeks after treatment, and the final levels of sex steroids found in treated fish at the time of harvesting are some 300 times lower than the levels found in wild fish (Solar and Donaldson, unpublished).

Table 4. Sterilization of fish treated with 17-alpha-methyltestosterone (extended from Hunter and Donaldson, 1983).

Species	Hormone Immersion	Treatment Dietary	Effect	References
Oncorhynchus keta & gorbuscha		50-100 mg/kg for 14 days at 1 year	Degeneration of spermatogonia	Yamazaki, 1972
O. Mykiss		15-60 mg/kg for 5 months from 1 month post hatching	12% sterile, 58% male	Jalabert et al., 1975
		50 mg/kg for 5 months from 1 month post hatch	Testes devoid of germ cells	Yamazaki, 1976
	Two X 2 hour of eyed eggs at 250 μg/L	3 mg/kg for 90 days from first feeding	17% sterile, 83% male	Johnson et al., 1978, 1979
		25 mg/kg for 110 days at 9 - 12°C	60% sterile	V.J. Bye in Hunter and Donaldson, 1983
	3-300 μg/L for 28 or 56 days		sterility and high mortality	van den Hurk and Slof, 1981
		0.5 mg/kg from June to November	gonadal inhibition	Billard et al., 1981
		25-50 mg/kg for 60-90 days from 30 days post hatch	79-90% sterile, paradoxical females obtained at higher doses	Solar et a., 1984
Salmo Salar	Two X 2 hour of eyed eggs at 250 μg/L	30 mg/kg for 120 days from alevin	no ovarian development observed	Simpson, 1975, 1976
		30 mg/kg for 120 days	sterile	Johnstone et al., 1978
		3 mg/kg for 90 days	male or sterile	Johnstone et al., 1978
		20 mg/kg for 600 degree days from first feeding	completely sterile	Johnstone, 1989
O. kisutch		1-50 mg/kg for 42 days at 8-9 months post hatch	reduction in spermatogonia	McBride and Fagerlund, 1973

con't....

Table 4. Continued:

Species	Hormone Immersion	Treatment Dietary	Effect	Reference
		0.2-10 mg/kg for 504 days starting 3-4 months post hatch	Gonadal inhibition at 10 mg/kg	Fagerlund and McBride, 1975
	Two X 2 hour of eyed egg and alevin at 25-400 µg/L	20 mg/kg for 70 days from first feeding	94-100% sterile	Goetz et al., 1979
	eyed eggs at 100-400 µg/L	10 mg/kg for 3 min.	70-94% sterile	Hunter et al., 1982a
	eyed eggs at 200 µg/L	10 mg/kg for 84 days	97.7% sterile	Donaldson and Hunter, 1982b
O. tshawytcha	5 X 2hour of alevin at 400 µg/L	40-80 mg/kg for 12 weeks	85-97% sterile	Solar et al., 1988
Tilapia macrochir		40 mg/kg for 2 months	100% sterile	Jalabert et al., 1974
Cyprinus carpio		400 ppm for 30 days from 1 day post hatch	61% sterile, 39% male	Basavaraja and Satyanarayana-Rao, 1988
		200-300 ppm for 50 days from 1 day post hatch	82-99.2% sterile, 0.8-18% male	Basavaraja and Satyanarayana-Rao, 1988
		300-400 ppm for 30 days from 1 day post hatch	86-98% sterile, 7-8% male, some females at 300 ppm	Manzoor and Satyanarayano-Rao, 1989
		200-300 ppm for 50 days from 1 day post hatch	82-99.2% sterile, 0.8-18% male	Basavaraja and Satyanarayana-Rao, 1988
		300-400 ppm for 30 days from 1 day post hatch	86-98% sterile, 2-8% male, some females at 300 ppm	Manzoor and Satyanarayana-Rao, 1989

C. Other Sterilization or Containment Techniques

Several other sterilization methods are described that have potential application to genetic containment of fish populations. These techniques are either still at the research or development phase, or have certain characteristics which currently limit their application to commercial aquaculture operations.

a) Surgery

Removal of gonadal tissue surgically is an extremely effective means to sterilize fish populations. In salmonids, this procedure has been used in rainbow trout (Wunder 1977), Atlantic salmon (Brown and Richards, 1979; Brown, 1982), and Pacific salmon (Robertson 1961; McBride et al. 1963). However, several difficulties exist with this method. Labour costs to implement this methodology commercially would be high. In addition, the removal of the gonads must be complete, particularly at the anterior end, or the gonadal tissue will undergo a process of compensatory growth and the fish may be still capable of producing gametes and sufficient hormones to induce final sexual maturation. Consequently, this procedure is most reliably conducted on reasonably large fish, and is not considered to be economically viable for most commercial situations.

b) Radiation treatment

Radiation treatment in the form of gamma rays has been shown to destroy germ cells, delay maturation, and induce functional sterility in a number of fish species (Bonham and Donaldson, 1972; Heshberger et al., 1978; Egami and Ijiri, 1979; Konno, 1980; Konno and Tashiro, 1982; Thorpe et al., 1987a and 1987b). With this technique, it appears to feasible to knock out germ cells and induce sterility, however the gonadal endocrine tissue will still induce final maturation (Egami and Hama-Furukawa, 1980; Thorpe et al. 1987b). Manion et al. (1988) have noted the partial effectiveness of cesium-137 irradiation on sterilization of released male sea lampreys as a control method to prevent reproduction between wild males and females.

c) Autoimmune sterility

Infertility in vertebrates has been accomplished by inducing an autoimmune response against several proteins involved in reproduction (Jones, 1982). In Atlantic salmon, Laird et al. (1978 and 1980) have obtained preliminary results suggesting that sterility could be induced by inoculation of smolts with adult gonadal tissue together with an adjuvant. A study in rainbow trout males also found that autoimmune sterility could be induced, but that the interstitial cells of the testis were still capable of producing sufficient androgen to induce sexual maturation (Secombes et al. 1987). For containment purposes such fish would behave much as do triploid males: they are reproductively contained but can still cause disruptions by interacting with fertile conspecifics on breeding grounds.

d) Chemosterilization

Chemosterilization methods have been investigated in fish, including the use of compounds which interfere with the normal action of sex steroids such as antiestrogens and antiandrogens (Schreck and Fowler, 1982), potent mutagens capable of disrupting germ-cell DNA (Tsoi, 1969), and through the use of gonadotropin antagonists such as methallibure (Donaldson, 1973; Flynn, 1973). The use of gonadotropin releasing hormone antagonists for sterilization in fish has not yet been reported. Sterilization of male sea lampreys for controlled reproduction has been achieved by treatment with bisazir (p,p-bis(1-aziridinyl)-N-methylphosphinothioic amide) by immersion (Hansen, 1981), the residues of which are cleared from the treated animals (Allen and Dawson, 1987).

e) Behavioral Control

Manipulation of certain aspects of salmonid behaviour have the potential to provide a degree of reproductive isolation. For example, the homing instinct is based in part upon recognizing particular olfactory clues present in their natal streams at the time of smolting (Quinn, 1984). Fish which have been exposed to characteristic water supplies, or chemicals such as morpholine (Cooper et al., 1976), during the fresh water phase of their life cycle will seek these same clues upon their return to spawn, migrating either to specific river systems or even to artificial sources of the chemicals dripped into salt water sites. The homing nature of salmon is exploited by the salmon ranching industry, and this characteristic may also be used for genetic containment by treating fish during their hatchery phase with distinct water supplies or chemicals, and, if an escape were to occur, use these same clues to assist in their recovery at the times they are expected to return to spawn. The use of triploid males (which are reproductively contained but will still return to natal rivers) in such programs could allow salmon ranching to proceed with little genetic risk to natural populations.

f) Hybrid sterility

The production of interspecific hybrids between different species of closely related fish can provide a simple method to cause sterility. Examples of interspecific hybridization in salmonids has been listed by Dangle et al. (1973) and reviewed by Chevassus (1979 and 1983). Clear examples of interspecific hybrid sterility have only been reported for crosses involving species from the genus *Salvelinus*, although ambiguous results also have been obtained for Atlantic salmon (*Salmo salar*) by brown trout (*Salmo trutta*) hybrids (see Chevassus, 1979). Although many of the possible interspecific crosses have been performed within the family Salmonidae, further work is required to evaluate the fertility of any surviving hybrids. Clear examples of hybrid sterility do exist in other fish species. In sunfish, male hybrids between the bluegill and green varieties are gametically sterile (sperm produced, but only inviable embryos obtained) in crosses to females of either parental species (Brunson and Robinette, 1987). Sterile interspecific hybrid grass carp have also been reported (Krasznai et al., 1984). In addition to sterility, the performance of any hybrids would need to be equivalent to, or better than, the parental species to be useful in a commercial setting. It may be possible to combine desirable qualities from several species into a "generic" salmonid strain suitable for the aquaculture industry.

g) Monosex Populations

In the case where no indigenous members of a cultured species are found in the local vicinity, and this species is not capable of forming fertile hybrids with any other local species, genetic containment may be achieved simply by using only one sex in production facilities. Any individuals which do escape would not be able to find suitable mates, and their potential for disruption would be limited to non reproductive activities. Broodstock populations of these species would obviously require both sexes, and should be appropriately contained in land-based facilities. Alternatively, monosex broodstock consisting of one genetic sex (but both phenotypic sexes maintained by artificial sex inversion) would require a lesser degree of containment since, in the case of an escape, their reproductive interaction would be limited to a single generation. Methods for the production and application of monosex populations of fish have been discussed by Donaldson (1986), Donaldson and Benfey (1987) and Dunham (1990).

h) Transgenic methods for inducing sterility

One of the major uses for sterilization techniques will be to allow containment of genetically engineered organisms. It is encouraging that this technology may also be able to provide a variety of methodologies to induce sterility in fish. The regulatory properties of genes are often contained within their upstream promoter regions, and these sequences can be moved in front of different structural genes to allow expression of particular proteins in discrete cell types. This characteristic has been utilized in transgenic mice to develop a procedure known as genetic ablation (Palmiter et al., 1987; Breitman et al., 1987; Behringer et al., 1988; Borrelli et al., 1989). These workers have shown that tissue-specific promoters can be used to drive the expression of a cytotoxic gene product (diptheria-toxin A chain or Herpes thymidine kinase) and specifically destroy discrete cell types. Infertility in male mice (Peschon et al., 1989; Braun et al., 1990) has been induced by driving the expression of the toxic protein in testicular cells with the use of a promoter from the mouse protamine gene (protamine is a sperm-specific protein). Similar approaches have been used to induce sterility in plant tissues by driving the expression of foreign ribonucleases in male reproductive tissues with anther-specific promoters (Mariani et al., 1990). Maclean and Penman (1990) have discussed this approach with reference to sterilization of fish strains.

A number of choices for sterilization of fish by genetic ablation can be envisaged, requiring only that the expression of the promoter being employed be restricted to tissues involved in reproduction. Unless the expression of the promoter can be controlled (for example induction of a critical protein at the correct reproductive stage to induce sterility in production fish, but not in broodstock), the sterility-inducing gene constructs would need to be kept in homozygous form in the sex for which they do not function, and then crossed into other stocks for production. This could prove to be a simple way of establishing transgenic populations for production that contain both sterilization constructs and inserts that confer a favourable phenotype (i.e. enhanced growth). For males, extension of the mouse work with protamine gene promoters is now feasible since these sequences are well characterized from salmonids (Gregory et al., 1982; Aiken et al., 1983).

Of importance for application of such methods to food-fish culture is the choice of ablation system employed with respect to environmental safety, human health, and public acceptance issues. For containment purposes, the stability of the transgenic modification would have to be assessed on a large scale before commercial implementation could be allowed. It should also be noted that other transgenic approaches could be employed to induce sterility in fish. Regulatory disruption of reproductive hormones in the hypothalamus-pituitary or pituitary-gonadal axes (for example, anti-sense constructs to gonadotropin or gonadotropin releasing hormone driven by inducible promoters) could also prevent normal gonadal maturation. Other mechanisms are also likely to be discovered. For example, in transgenic mice, Hammer et al. (1984) have noted that high level expression of growth hormone can induce female infertility, and this seems to occur by disrupting the required postcopulatory release of prolactin (Bartke et al., 1988).

SUMMARY

The development of genetically altered strains of fish for aquaculture requires the co-implementation of containment methodologies that can effectively block reproductive interaction with wild populations. Genetically altered fish stocks may possess either rearrangement (with respect to DNA sequence, or population frequency of alleles) of their natural genetic information, or may contain new genetic information derived from hybridization or transgenesis. Although it is the phenotype that ultimately interacts with the environment, carefull consideration of the altered genotype is also required to evaluate the long-term potential for phenotypic change in the future. Control of genetically altered fish can be achieved by a combination of physical and biological containment methodologies. Physical containment measures are highly dependent on the form of culture, but, in general, should apply the most stringent engineering and management systems available. Biological containment is best achieved through sterilization procedures that prevent the reproduction of escaped cultured fish with their wild counterparts. Although sterility in fish can be induced by a variety of agents, induction of triploidy and treatment with high doses of androgen are currently the most effective large-scale measures available. The effective implementation of containment measures will require monitoring of aquaculture facilities for compliance with regulations, and systematic surveying of the natural environment for the existence and impact of escaped fish.

REFERENCES

Aiken, J.M., McKenzie, D., Zhao, H.-Z., States, J.C. and Dixon, G.H. (1983) Sequence homology in the protamine gene family of rainbow trout. Nucleic Acids Res. 11: 4907-4922.

Allen, J.L. and Dawson, V.K. (1987) Elimination of super(14)C-bisazir residues in adult sea lamprey (*Petromyzon marinus*). Tech. Rep. GLFC. No. 50, pp. 9-17.

Altukhov, Yu.P. (1981) The stock concept from the viewpoint of population genetics. Can. J. Fish. Aquat. Sci. 12: 1523-1538.

Altukhov, Yu.P. and Salmenkova, E.A. (1987) Stock transfer relative to natural organization, management and conservation of fish population. In: Population genetics and fishery management. N. Ryman and F. Utter (eds.) University of Washington Press, Seattle, WA. pp. 333-343.

Bain, M.B., Webb, D.H., Tangedal, M.D. and Mangum, L.N. (1990) Movements and habitat use by grass carp in a large mainstream reservoir. Trans. Am. Fish. Soc. 19: 553-561.

Baker, I.J., Solar, I.I. and Donaldson, E.M. (1988) Masculinization of chinook salmon (*Oncorhynchus tshawytscha*) by immersion treatments using 17 α-methyltestosterone around the time of hatching. Aquaculture. 72: 359-367.

Baker, I.J., Solar, I.I., Mulji, K., Donaldson, E.M., Hunter, G.A. and Stone, E.T. (1989) Coded wire tag recoveries from the second release of sterile coho salmon (*O. kisutch*) into the marine environment. Can. Data Rep. Fish.Aquat. Sci. No. 775. 21 pp.

Bams, R.A. (1976) Survival and propensity for homing as affected by presence or absence of locally adapted paternal genes in two transplanted populations of pink salmon (*Oncorhynchus gorbuscha*). J. Fish. Res. Board Can. 33: 2716-2725.

Bartke, A., Steger, R.W., Hodges, S.L., Parkening, T.A., Collin, T.J., Yun, J.S. and Wagner, T.E. (1988) Infertility in transgenic female mice with human growth hormone expression: evidence for luteal failure. J. Exp. Biol. 248: 121-124.

Basavaraja, N. and Satyanarayana Rao, G.P. (1988) Induction of sterility in the common carp, *Cyprinus carpio* (Linn.) using 17alpha-methyltestosterone. Indian J. Anim. Sci. 58: 1248-1251.

Behringer, R. R., Mathews, L. S., Palmiter, R. D. and Brinster, R. L., (1988) Dwarf mice produced by genetic ablation of growth-hormone producing cells. Genes and Development. 2: 453-461.

Benfey, T.J. and Sutterlin, A.M. (1984) Growth and gonadal development in triploid landlocked Atlantic salmon (*Salmo salar*). Can. J. Fish. Aquat. Sci. 41: 1387-1392.

Benfey, T.J., Solar, I.I., De Jong, G. and Donaldson, E.M. (1986) Flow-cytometric confirmation of aneuploidy in sperm from triploid rainbow trout. Trans. Am. Fish. Soc. 115: 838-840.

Benfey, T.J. and Donaldson, E.M. (1988) Triploidy in the culture of Pacific salmon. In: Proc. Aquaculture International Congress (Vancouver, Canada: Sept. 6-9, 1988), pp. 549-554. British Columbia Pavilion Corp., Vancouver.

Benfey, T.J., Bosa, P.G., Richardson, N.L. and Donaldson, E.M. (1988) Effectiveness of a commercial-scale pressure shocking device for producing triploid salmonids. Aquac. Engineer. 7: 147-154.

Benfey, T.J. (1989) A bibliography of triploid fish, 1943 to 1988. Can. Data Rep. Fish. Aquat. Sci. No. 1682. 33 pp.

Benfey, T.J. Dye, H.M., Solar, I.I. and Donaldson, E.M. (1989) The growth and reproductive endocrinology of adult triploid Pacific salmonids. Fish Physiol. Biochem. 6: 113-120.

Bidwell, C. A., Chrisman, C. L. and Libey, G. S., (1985) Polyploidy induced by heat shock in channel catfish. Aquaculture. 51: 25-32.

Billard, R., Breton, B. and Richard, M. (1981) On the inhibitory effect of some steroids on spermatogenesis in adult rainbow trout (*Salmo gairdneri*). Can. J. Zool. 59: 1479-1504.

Bonham, K. and Donaldson, L.R. (1972) Sex ratios and retardation of gonadal development in chronically gamma-irradiated chinook salmon smolts. Trans. Am. Fish. Soc. 101: 428-434.

Borrelli, E., Heyman, R.A., Arias, C., Sawchenko, P.E. and Evans, R.M. (1989) Transgenic mice with inducible dwarfism. Nature. 339: 538-541.

Boulanger, Y. (1991) Performance comparison of all-female, diploid and triploid brook trout. In: Proc. Atlantic Canada Workshop on Methods for the Production of non-maturing salmonids: February 19-21, 1991. Dartmouth, Nova Scotia, (Pepper, V.A., ed.). Can. Tech. Rep. Fish. Aquat. Sci. No. 1789, 111-121.

Braun, R.E., Lo, D., Pinkert, C.A., Widera, G., Flavell, R.A., Palmiter, R.D. and Brinster, R.L. (1990) Infertility in male transgenic mice: disruption of sperm development by HSV-tk expression in postmeiotic germ cells. Biol. Reprod. 43: 684-693.

Breitman, M.L., Clapoff, S., Rossant, J., Tsui, L-C., Glode, M., Maxwell, I.H. and Bernstein, A. (1987) Genetic ablation: targeted expression of a toxin gene causes microphthalmia in transgenic mice. Science. 238: 1563-1565.

Brown, L.A. and Richards, R.H. (1979) Surgical gonadectomy of fish: a technique for veterinary surgeons. Vet. Rec. 104: 215.

Brown, L.A. (1982) Surgical gonadectomy of salmonid fish and subsequent prevention of diseases associated with sexual maturation. Proc. R. Soc. Edinburgh, Sect. B. 81: 211-219.

Brunson, M.W. and Robinette, H.R. (1987) Reproductive isolation between a hybrid sunfish and its parental species. Prog. Fish-Cult. 49: 296-298.

Bye, V.J. and Lincoln, R.F. (1986) Commercial methods for the control of sexual maturation in rainbow trout (*Salmo gairdneri R.*). Aquaculture. 57: 299-309.

Chevassus, B. (1979) Hybridization in salmonids: results and perspectives. Aquaculture. 17: 113-128.

Chevassus, B. (1983) Hybridization in fish. Aquaculture. 33: 245-262.

Chevassus, B., Devaux, A., Chourrout, D. and Jalabert, B. (1988) Production of YY rainbow trout males by self-fertilization of induced hermaphrodites. J. Hered. 79: 89-92.

Chilcote, M.W., Leider, S.A. and Loch, J.J. (1986) Differential reproductive success of hatchery and wild summer-run steelhead under natural conditions. Trans. Am. Fish. Soc. 115: 726-735.

Chourrout, D. (1982) Tetraploidy induced by heat shocks in the rainbow trout (*Salmo gairdneri* R.). Reprod. Nutr. Dev. 22: 569-574.

Chourrout, D. (1984) Pressure-induced retention of second polar body and suppression of first cleavage in rainbow trout: production of all-triploids, all-tetraploids, and heterozygous and homozygous diploid gynogenetics. Aquaculture 36: 111-126.

Chourrout, D., Chevassus, B., Krieg, F., Happe, A., Burger, G. and Renard, P. (1986) Production of second generation triploid and tetraploid rainbow trout by mating tetraploid males and diploid females - potential of tetraploid fish. Theor. Appl. Genet. 72: 193-206.

Chourrout, D. (1987) Genetic manipulations in fish: review of methods. In: Selection, Hybridization, and Genetic Engineering in Aquaculture, Vol. II (Tiews, K., ed.), pp. 111-126. Heenemann Verlags. mbH, Berlin.

Chourrout, D., and Nakayama, I. (1987) Chromosome studies of progenies of tetraploid female rainbow trout. Theor. Appl. Genet. 74: 687-692.

Cooper, J.C., Scholz, A.T. and Hasler, A.D. (1976) Experimental confirmation of the olfactory hypothesis with artificially imprinted homing coho salmon (*Oncorhynchus kisutch*). J. Fish. Res. Board Can. 33: 703-710.

Dangel, J.R., Macy, P.T., and Withler, P.C. (1973) Annotated bibliography of interspecific hybridization of fishes of the subfamily Salmoninae. NOAA Tech. Mem., NMFS NWFC. 1:48.

Department of Fisheries and Oceans. (1991) International Symposium on Biological Interactions of enhanced and wild salmonids, June 17-20, 1991. Nanaimo, B.C. Book of Abstracts. 80 pp.

Devlin, R.H., McNeil, K.B., Groves, T.D.D. and Donaldson, E.M. (1991) Isolation of a Y-chromosome DNA probe capable of determining genetic sex in chinook salmon (*Oncorhynchus tshawytscha*). Can. J. Fish. Aquat. Sci. (in press).

Don, J. and Avtalion, R. R., (1988) Comparative study on the induction of triploidy in tilapias, using cold- and heat-shock techniques. *J. Fish Biol.* 32: 665-672.

Donaldson, E.M. (1973) Reproductive endocrinology of fishes. Am. Zool. 13: 909-927.

Donaldson, E.M. and Hunter, G.A. (1982a) Sex control in fish with particular reference to salmonids. Can. J. Fish. Aquat. Sci. 39: 99-110.

Donaldson, E.M. and Hunter, G.A. (1982b) The ocean release and contribution to the fishery of all-female and sterile groups of coho salmon (*Oncorhynchus kisutch*). Proc. Int. Symp. Reprod. Physiol. Fish. 1982, p. 78.

Donaldson, E.M. (1986) The integrated development and application of controlled reproduction techniques in Pacific salmonid aquaculture. Fish Physiol. Biochem. 2: 9-24.

Donaldson, E.M. and Benfey, T.J. (1987) Current status of induced sex manipulation. In: Proc. Third Int. Symp. on the Reproductive Physiology of Fish (Idler, D.R., Crim, L.W. and Walsh, J.M., eds.), pp. 108-119. Memorial Univ. of Nfld., St. John's.

Donaldson, E.M., Piferrer, F., Solar, I.I. and Devlin, R.H. (1991a) Studies on hormonal sterilization and monosex technologies for salmonids at the West Vancouver Laboratory. Can. Tech. Rep. Fish. Aquat. Sci. No. 1789: 37-45.

Donaldson, E.M., Devlin, R.H., Solar, I.I., and Piferrer, F. (1991b) The reproductive containment of genetically altered salmonids. In: NATO Advanced Studies Institute Series. (J.G. Cloud and G.H. Thorgaard, eds.). Plenum Publishing Corp. (in press).

Dunham, R.A. (1990) Production and use of monosex or sterile fishes in aquaculture. Reviews in Aquatic Sciences. 2: 1-17.

Eckstein, B. and Spira, M. (1965) Effect of sex hormones on the gonadal differentiation in a cichlid, *Tilapia aurea*. Biol. Bull. (Woods Hole, Mass.) 129: 482-489.

Egami, N. and Ijiri, K.I. (1979) Effects of irradiation on germ cells and embryonic development in teleosts. Int. Rev. Cytol. 59: 195-248.

Egidius, E., Hansen, L.P., Jonsson, B. and Naevdal, G. (1991) Mutual impact of wild and cultured Atlantic salmon in Norway. J. Cons. int. Explor. Mer, 47: 404-410.

Ellis, A.E. (1981) A field trial of a method to induce auto-immune castration in *Salmo salar*. Int. Coun. Explor. Sea. CM 1981/F:37. 10 pp.

Fagerlund, U.H.M. and McBride, J.R. (1975) Growth increments and some flesh and gonad characteristics of juvenile coho salmon receiving diets supplemented with 17α-methyltestosterone. J. Fish Biol. 7: 305-314.

Federal Register. (1990) Research proposal on transgenic fish: publication of environmental assessement. Federal Register. 55: 5752-5757.

Ferguson, M.M., Danzmann, R.G., and Allendorf, F.W. (1988) Developmental success of hybrids between two taxa of salmonid fishes with moderate structural gene divergence. Can. J. Zool. 66: 1389-1395.

Ferguson, M.M. (1990) The genetic impact of introduced fishes on native species. Can. J. Zool. 68: 1053-1057.

Fletcher, G.L., Shears, M.A., King, M.J., Davies, P.L. and Hew, C.L. (1988) Evidence for antifreeze protein gene transfer in fish: The first step toward a more freeze-resistant salmon. Can. J. Fish. Aquat. Sci. 45: 352-357.

Flynn, M.B. (1973) The effect of methallibure and a constant 12- hours-light : 12-hours-dark photoperiod on the gonadal maturation of pink salmon (*Oncorhynchus gorbuscha*). Masters Thesis Def. UBC. Vancouver B.C., pp. 70

Fraser, J.M. (1981) Comparative survival and growth of planted, wild, hybrid, and domestic strains of brook trout (*Salvelinus fontinalis*) in Ontario lakes. Can. J. Fish. Aquat. Sci. 38: 1672-1684.

Furevik, D., Rabben, H., Mikkelsen, K.O. and Fosseidengen, J.E. (1990) Migratory patterns of escaped farmraised Atlantic salmon. Int. Council Explor. Sea. Maricul. Comm. C.M. 1990/F:55.

Gausen, D. and Moen, V. (1991) Large-scale escapes of farmed Atlantic salmon (*Salmo salar*) into Norwegian rivers threaten natural populations. Can. J. Fish. Aquat. Sci. 48: 426-428.

Goetz, F.W., Donaldson, E.M., Hunter, G.A. and Dye, H.M. (1979) Effects of estradiol-17ß and 17α-methyltestosterone on gonadal differentiation in the coho salmon, *Oncorhynchus kisutch*. Aquaculture. 17: 267-278.

Goubau, S., Silversides, D.W., Gonzalez, A., Laarveld, B., Mapletoft, R.J. and Murphy, B.D. (1989) Immunization of sheep against modified peptides of gonadotropin releasing hormone conjugated to carriers. Domest. Anim. Endocrinol. 6: 339-347.

Gregor, S.P., Dillon, N.O. and Butterworth, P.H.W. (1982) The localisation of the 5' termini of in vivo and in vitro transcripts of a cloned rainbow trout protamine gene. Nucleic Acids Res. 10: 7581-7592.

Guoxiong, C., Solar, I.I. and Donaldson, E.M. (1989) Comparison of heat and hydrostatic pressure shocks to induce triploidy in steelhead trout (*Oncorhynchus mykiss*). Can. Tech. Rep. Fish. Aquat. Sci. No. 1718, 11 pp.

Hammer, R.E., Palmiter, R.D. and Brinster, R.L. (1984) Partial correction of a murine heredity growth disorder by germ-line incorporation of a new gene. Nature. 311: 65.

Hanson, L.H. (1981) Sterilization of sea lampreys (*Petromyzon marinus*) by immersion in an aqueous solution of Bisazir. Can. J. Fish. Aquat. Sci. 38: 1285-1289

Hansen, L.P., Doving, K.B. and Jonsson, B. (1987) Migration of farmed adult Atlantic salmon with and without olfactory sense, released on the Norwegian coast. J. Fish Biol. 30: 713-721.

Hershberger, W.K., Bonham, K. and Donaldson, L.R. (1978) Chronic exposure of chinook salmon eggs and alevins to gamma irradiation: effects on their return to freshwater as adults. Trans. Am. Fish. Soc. 107: 622-631.

Hirose, K. and Hibiya, T. (1968) Physiological studies on growth-promoting effect of protein-anabolic steroids on fish. II. Effects of 4-chlorotestosterone acetate on rainbow trout. Bull. Jpn. Soc. Sci. Fish. 34: 473-481.

Hunter, G.A., Donaldson, E.M., Goetz, F.W., and Edgell, P.R. (1982) Production of all female and sterile groups of coho salmon (*Oncorhynchus kisutch*) and experimental evidence for male heterogamety. Trans. Am. Fish. Soc. 111: 367-372.

Hunter, G.A. and Donaldson, E.M. (1983) Hormonal sex control and its application to fish culture. In: Fish Physiology, Vol. IX, Reproduction, Part B, Behavior and Fertility Control (W.S. Hoar, D.J. Randall and E.M. Donaldson, eds.). Chap. 5. p. 223-303. Academic Press, New York, N.Y.

Hunter, G.A., Donaldson, E.M., Stoes, J., and Baker, I. (1983) Production of monosex female groups of chinook salmon (*Oncorhynchus kisutch*) by the fertilization of normal ova with sperm from sex reversed females. Aquaculture. 33: 355-364.

Jalabert, B., Moreau, J., Planquette, P. and Billard, R. (1974) Determinisme du sexe chez *Tilapia macrochir* et *Tilapia nilotica*: Action de la methyltestosterone dans l'alimentation des alevins sur la differenciation sexuelle, proportion des sexes dans la descendance des males "inverses". Ann. Biol. Anim., Biophys. 14(4-B): 729-739.

Jalabert, B., Billard, R. and Chevassus, B. (1975) Preliminary experiments on sex control in trout: Production of sterile fishes an simultaneous self-fertilization hermaphrodites. Ann Biol. Anim., Biophys. 15: 19-28.

Johnstone, R., MacIntosh, D.J. and Wright, R.S. (1983) Elimination of orally administered 17 alpha-methyltestosterone by *Oreochromis mossambicus* (tilapia) and *Salmo gairdneri* (rainbow trout) juveniles. Aquaculture. 35: 249-257.

Johnstone, R. and Youngson, A.F. (1984) The progeny of sex inverted female Atlantic salmon (*Salmo salar L.*). Aquaculture. 37: 179-182.

Johnstone, R. (1985) Induction of triploidy in Atlantic salmon by heat shock. Aquaculture 49: 133-139.

Johnstone, R. (1989) Maturity control in Atlantic salmon. A review of the current status of research in Scotland. In: M. Carrillo, S. Zanuy and S. Planas, Compilers, XIth Int. Symp. Comp. Endocrinol., Proc. of the Satellite Symp. on Applications of Comp. Endocrinol. to Fish Culture, Almunecar, Spain. May 22-23, 1989. p. 89-94.

Johnstone, R., Knott, R.M., Macdonald, A.G. and Walsingham, M.V. (1989) Triploidy induction in recently fertilized Atlantic salmon ova using anaesthetics. Aquaculture. 78: 229-236.

Johnstone, R., McLay, H.A. and Walsingham, M.V. (1991) Production and performance of triploid Atlantic salmon in Scotland. Can. Tech. Rep. Fish. Aquat. Sci. No. 1789, 15-36.

Jones, W.R. (1982) Immunological fertility regulation. Blackwell Sci. Publ., Oxford, London, Edinburgh, Boston, Melbourne.

Jungalwalla, P.J. (1991) Production of non-maturing Atlantic salmon in Tasmania. Can. Tech. Rep. of Fish. Aquat. Sci. No. 1789, 47-71.

Kapuscinski, A.R. and Hallerman, E.M. (1990) Transgenic fish and public policy: anticipating environmental impacts of transgenic fish. Fisheries. 15:2-11.

Konno, K. (1980) Effects of gamma-irradiation on the gonads of the rainbow trout, *Salmo gairdneri irideus*, during embryonic stages. In: Radiation effects on aquatic organisms. N. Egami (ed). Japan Soc. Sci. konPress, Tokyo/Univ. Park Press, Baltimore. pp. 129-133.

Konno, K. and Tashiro, F. (1982) The sterility of rainbow trout (*Salmo gairdneri*) irradiated with cobalt-60 gamma rays. J. Tokyo Univ. Fish. 68: 75-80.

Krasznai, Z., Marian, T., Buris, L. and Ditroi, F. (1984) Production of sterile hybrid grass carp (*Ctenopharyngodon idella* Val. x *Aristichthys nobilis* Rich.) for weed control. Aquacult. Hung. 4: 33-38.

Laird, L.M., Ellis, A.E., Wilson, A.R. and Holliday, F.G.T. (1978) The development of the gonadal and immune systems in the Atlantic salmon (*Salmo salar L.*) and a consideration of the possibility of inducing autoimmune destruction of the testis. Ann. Biol. Anim., Biochim., Biophys. 18: 1101-1106.

Laird, L.M., Wilson, A.R. and Holliday, F.G.T. (1980) Field trials of a method of induction of autoimmune gonad rejection in Atlantic salmon (*Salmo salar L.*). Reprod. Nutr. Dev. 20: 1781-1788.

Loverich, G.F. (1991) A discussion of two offshore experiments with ocean spar net pens. Aquaculture Association of Canada, 8th Annual meeting. Technical program and abstracts, June 10-13, 1991. St. Andrews, New Brunswick, Canada, pp. 20.

MacLean, J.A., Evans, D.O., Martin, N.V., and Desjardine, R.L. (1981) Survival, growth, spawning distribution, and movements of introduced and native lake trout (*Salvelinus namaycush*) in two inland Ontario lakes. Can. J. Fish. Aquat. Sci. 38: 1685-1700.

Maclean, N., and Penman, D. (1990) The application of gene manipulation to aquaculture. Aquaculture. 85: 1-20.

McBride, J.R., Fagerlund, U.H.M., Smith, M. and Tomlinson, N. (1963) Resumption of feeding by and survival of adult sockeye salmon (*Oncorhynchus nerka*) following advanced gonad development. J. Fish. Res. Board Can. 20: 95-100.

McBride, J.R. and Fagerlund, U.H.M. (1973) The use of 17 α-methyltestosterone for promoting weight increases in juvenile Pacific salmon. J. Fish. Res. Board Can. 30: 1099-1104.

McCarter, N.H. (1988) Verification of the production of triploid grass carp (*Ctenopharyngodon idella*) with hydrostatic pressure. N.Z. J. Mar. Freshwat. Res. 22: 501-505.

McLean, E., Sadar, M.D., Devlin, R.H., Donaldson, E.M. and Souza, L.M. (1991) Promotion of growth in diploid and triploid coho salmon following parenteral delivery of a recombinant porcine somatotropin. Aquat. Liv. Resources. (in press).

Manion, P.J. Hanson, L.H. and Fodale, M.F. (1988) Sterilizing effect of cesium-137 irradiation on male sea lampreys released in the Big Garlic River, Michigan. Tech. Rep. GLFC. No. 53, pp. 1-7.

Manzoor Ali, P.K.M. and Satyanarayana Rao, G.P. (1989) Growth improvement in carp, *Cyprinus carpio* (Linnaeus), sterilized with 17alpha-methyltestosterone. Aquaculture. 76: 157-167.

Mariani, C., De Beuckeleer, M., Truettner, J., Leemans, J. and Goldberg, R.B. (1990) Induction of male sterility in plants by a chimaeric ribonuclease gene. Nature. 347: 737-741.

Moyle, P.B. (1986) Fish introduction into North America: patterns and ecological impacts. In: Ecology of biological invasions of North America and Hawaii. Ecological Studies 58 (Mooney, H.A. and Drake, J.A., eds.), pp. 27-40. Springer-Verlag, New York, New York, USA.

Myers, J.M., Hershberger, W.W., and Iwamoto, R.N. (1986) The induction of tetraploidy in salmonids. J. World Mariculture Soc. 17: 1-4.

NASCO. (1990) Report on the Norwegian Meeting on impacts of aquaculture on wild stocks. North Atlantic Salmon Conservation Organization. Paper CLN(90)28: 9 pp.

Nickelson, T.E., Solazzi, M.F. and Johnson, S.L. (1986) Use of hatchery coho salmon (*Oncorhynchus kisutch*) to rebuild wild populations in Oregon coastal streams. Can. J. Fish. Aquat. Sci. 43: 2443-2449.

Okada, H., Matsumoto, H. and Yamazaki, F. (1979) Functional masculinization of genetic females in rainbow trout. Bull. Jpn. Soc. Sci. Fish. 45: 413-419.

Palmiter, R.D., Behringer, R.R., Quaife, C.J., Maxwell, F., Maxwell, I.H. and Brinster, R.L. (1987) Cell lineage ablation in transgenic mice by cell-specific expression of a toxin gene. Cell. 50: 435-443.

Parsons, J.E. and Thorgaard, G.H. (1985) Production of androgenetic diploid rainbow trout. J. Hered. 76: 177-181.

Pepper, V.A. (1991) Proceedings of the Atlantic Canada Workshop on methods for the production of non-maturing salmonids. Can. Tech. Rep. Fish. Aquat. Sci. No. 1789, 152 pp.

Peschon, J.J. (1989) Expression of mouse protamine 1 genes in transgenic mice. Ann. NY Acad. Sci. 564: 186-197.

Philipp, D.P. (1990) Stock transfer of largemouth bass: long-term consequences for natural populations. In: Ecological and genetic implications of fish introductions (Fin), Abstracts, May 17-19, 1990. Great Lakes Institute, Univeristy of Windsor, p. 10.

Piferrer, F. and Donaldson, E.M. (1991) Dosage-dependent differences in the effect of of aromatizable and non-aromatizable androgens on the resulting phenotype of coho salmon (*Oncorhynchus kisutch*). Fish Physiol. Biochem. 9: 145-150.

Purdom, C.E. (1986) Genetic techniques for control of sexuality in fish farming. Fish Physiol. Biochem. 2: 3-8.

Quinn, T.P. (1984) Homing and straying in Pacific salmon. In: Mechanisms of migration in fishes, (McCleave, J.D., Arnold, G.P., Dodson, J.J. and Neill, W.H., eds.), pp. 357-362. Plenum Press, New York, NY.

Reddy, P.V.G.K., Kowtel, G.V. and Tantia, M.S., (1990) Preliminary observations on induced polyploidy in Indian major carps, *Labeo rohita* (Ham.) and *Catla catla* (Ham.). Aquaculture. 87: 279-287.

Reisenbirchler, R.R. and McIntyre, J.D. (1977) Genetic differences in growth and survival of juvenile hatchery and wild steelhead trout. J. Fish. Res. Board Can. 34: 123-128.

Robertson, O.H. (1961) Prolongation of the life span of kokanee salmon (*Oncorhynchus nerka kennerlyi*) by castration before beginning of gonad development. Proc. Natl. Acad. Sci. 47: 609-621.

Rosenthal, H. (1988) Principles and practice in quarantine operation. In: Efficiency in aquaculture production: disease control. Proc. 3rd Int. Conf. on Aquafarming Acquacoltura'86, Verona-Italy, October 9-10, 1986, (Grimaldi, E. and Rosenthal, H., eds.), pp. 179-205.

Schreck, C.B. and Fowler, L.G. (1982) Growth and reproductive development in fall chinook salmon: effects of sex hormones and their antagonists. Aquaculture. 36: 253-263.

Secombes, C.J., Laird, L.M. and Priede, I.G. (1987) Immunological approches to control maturation in fish. II. A review of the autoimmune approach. Aquaculture 60: 287-302.

Simpson, T.H. (1975-76) Endocrine aspects of salmonid culture. Proc. R. Soc. Edinburgh (B). 17: 241-252.

Skaala, O., Dahle, D., Dorstad, K.E., and Naevdal, G. (1990) Interactions between natural and farmed fish populations: information from genetic markers. J. Fish Biol. 36: 449-460.

Skaala, O., Jorstad, K.E., and Borgstrom, R. (1991) Genetic impact on wild populations from fish farming: experiments with genetically tagged brown trout (*Salmo trutta L.*). In: Abstracts from the International Symposium on Biological Interactions of Enhanced and Wild Salmonids. June 17-20, 1991. Nanaimo British Columbia, Canada p. 22.

Solar, I.I., Baker, I.J. and Donaldson, E.M. (1988) Effect of immersion and dietary treatment with 17α-methyltestosterone in the gonadal development of chinook salmon. First Int. Symp. Fish Endocrinol., Edmonton, Alta., June 12-17, 1988. Abstr. p. 38.

Solar, I.I., Donaldson, E.M. and Hunter, G.A. (1984) Induction of triploidy in rainbow trout (*Salmo gairdneri* Richardson) by heat shock, and investigation of early growth. Aquaculture. 42: 57-67.

Solar, I.I., Baker, I.J., Donaldson, E.M., Hunter, G.A. and Stone, E.T. (1986) Coded wire tag recoveries from the first release of all-female and sterile groups of coho salmon (*O. kisutch*) into the marine environment. Can. Data Rep. Fish. Aquat. Sci. No. 609. 29 pp.

Sutterlin, A.M. and Collier, C. (1991) Some observations of the commercial use of triploid rainbow trout and Atlantic salmon in Newfoundland, Canada. In: Proc. Atlantic Canada Workshop on Methods for the Production of non-maturing salmonids: February 19-21, 1991. Dartmouth, Nova Scotia, V. A. Pepper (ed.). Can. Tech. Rep. Fish. Aquat. Sci. No. 1789, 89-96.

Taggart, J.B. and Ferguson, A. (1986) Electrophoretic evaluation of a implemental stocking programme for brown trout, *Salmo trutta*. Aquacult. Fish. Manage. 17: 155-162.

Teskeredzic, E., Donaldson, E.M., Teskeredzic, Z., McLean, E. and Solar, I.I. (1991) Comparison of heat and heat-electro shocks to induce triploidy in coho salmon (*Oncorhynchus kisutch*). Can. Tech. Rep. Fish. Aquat. Sci. 1785, 7 pp.

Thorgaard, G.H., Jazwin, M.E. and Stier, A.R. (1981) Polyploidy induced by heat shock in the rainbow trout. Trans. Am. Fish. Soc. 110: 546-550.

Thorgaard, G.H. (1983) Chromosome set manipulation and sex control in fish. In: Fish Physiology, Vol. IX, Reproduction, Part B, Behavior and Fertility Control. Hoar, W.S., Randall, D.J. and Donaldson, E.M. (eds.). Academic Press, New York. pp. 405-434.

Thorgaard, G.H. (1986) Ploidy manipulation and performance. Aquaculture. 57: 57-64.

Thorpe, J.E., Talbot, C. and Miles, M.S. (1987a) Irradiation of Atlantic salmon eggs to overcome early maturity when selecting for high growth rate. Selection, Hybridization, and Genetic Engineering in Aquaculture, Vol. I, (Tiews, K., ed.), p. 361-374. Heenemann Verlags mbH, Berlin.

Thorpe, J.E., Wright, R.S., Talbot, C. and Miles, M.S. (1987b) Secondary sexual characteristics developed by 60-Co sterilised Atlantic salmon. Proc. Third Int. Symp. on the Reproductive Physiology of Fish, St. John's, Nfld. Aug. 2-7, 1987. p. 139.

Tiedje, J.M., Colwell, R.K., Grossman, Y.L., Hodson, R.E., Lenski, R.E., Mack, R.N. and Regal, P.J. (1989) The planned introduction of genetically engineered organisms: ecological considerations and recommendations. Ecology. 70: 298-315.

Tsoi, R.M. (1969) The effect of nitrosomethyl urea and dimethylsulphate on sperm of the rainbow trout (*Salmo irideus* Gibb.) and the peled (*Coregonus peled* Gmel.). Dokl. Biol. Sci. USSR 189: 849-852.

Ueda, T., Kobayashi, M. and Sato, 4. (1986) Triploid rainbow trouts induced by polyethylene glycol. Proc. Japan Acad. 62B: 161-164.

Ueda, T., Sato, R., Iwata, M., Komaru, A. and Kobayashi, J. (1991) The viable 3.5n trouts produced between diploid females and allotriploid males. Jpn. J. Genet. 66: 71-75.

Utter, F.M., Johnson, O.W., Thorgaard, G.H. and Rabinovitch, P.S.(1983) Measurement and potential applications of induced triploidy in Pacific salmon. Aquaculture 35: 125-135.

van den Hurk, R. and Slof, G.H. (1981) A morphological and experimental study of gonadal sex differentiation in the rainbow trout, *Salmon gairdneri*. Cell Tissue Res. 218: 487-497.

Welcomme, R.L. (1981) Register of international transfers of inland fish species. Food and Agriculture Organization Fisheries Technical Paper No. 213.

Welcomme, R.L. (1984) International transfers of inland fish species. In: Distribution, biology, and management of exotic fishes. Edited by W.R. Courtney and J.R. Stauffer. The Johns Hopkins University Press, Baltimore. pp. 22-40.

Willinsky, M.D., Robson, D.R., Vangool,W.J., Fournier,R.A., and Allen, J.H. (1991) Design of a spherical submersible self-cleaning aquaculture system for exposed sites. Aquaculture Association of Canada. 8th Annual meeting. Technical program and abstracts. June 10-13, 1991. St. Andrews, New Brunswich, Canada.

Wu, W., Li, C., Liu, G., Xu, D.,Liu, C., Xie, J., Shan, C. (1988) Studies on tetraploid hybrid between red common carp (*Cyprinus carpio*) and grass carp (*Ctenopharyngodon idellus*) and its backcross triploid. ACTA Hydrobiol. Sin. 12:355-363.

Wunder, W. (1977) Experimentelle untersuchungen zur frage der regeneration der maennlichen und weiblichen keimdruesen (gonaden) be der regenbozen forelle (*Salmo irideus* Gibb.). Zool. Anz. 198: 245-262.

Yamazaki, F. (1972) Effects of methyltestosterone on the skin and the gonads of salmonids. Gen. Comp. Endocrinol., Suppl. 3: 741-750.

Yamazaki, F. (1983) Sex control and manipulation in fish. Aquaculture 33: 329-354.

Zhang, P., Hayat, M., Joyce, C., Gonzalez-Villasenor, L.I., Lin, C.M., Dunham, R.A., Chen, T.T. and Powers, D.A. (1990) Gene transfer, expression and inheritance of pRSV-rainbow trout-GH cDNA in the common carp, *Cyprinus carpio* (Linnaeus). Mol. Reprod. Dev. 25: 3-13.

SUBJECT INDEX

D

K

L

M

N

O

P

Q

R

www.ingramcontent.com/pod-product-compliance
Ingram Content Group UK Ltd.
Pitfield, Milton Keynes, MK11 3LW, UK
UKHW020358150926
8624IPUK00004B/227

9 789810 209971